AF454089

Sustainable Development of Natural Resources and Wildlife Conservation

ABOUT THE EDITOR

Dr. Ashwani Kumar Dubey is President of Environment and Social Welfare Society, Khajuraho and Editor-in-Chief of "International Journal of Global Science Research, India". He is currently working as Guest Lecturer in Department of Zoology, Government Maharaja Post Graduate College, Chhatarpur, Madhya Pradesh.

He did M. Sc. with specialization of Ichthyology in 1991 from Awadhesh Pratap Singh University, Rewa, Madhya Pradesh. He obtained his Ph.D. Degree in 1995 entitled "Responses of antioxidants, lipid-protein interactions and lipid peroxidation in *Heteropneustes fossilis* to oxidative damage exposure" from School of Studies in Zoology, Vikram University, Ujjain, Madhya Pradesh, India. His research field was Biochemistry, Free Radical Biology, Toxicology and Stress Monitoring. He has published several research papers in Foreign Journals including Pergamon press United Kingdom to his credit and published many popular science articles for the student welfare.

Dr. Dubey is currently working on Impact of environmental degradation, global health, natural resource, wildlife conservation and stress monitoring. His personal interests including Reading, Writing, Traveling and Photography.

Sustainable Development of Natural Resources and Wildlife Conservation

Editor

Dr. Ashwani Kumar Dubey

2015

Daya Publishing House®

A Division of

Astral International (P) Ltd

New Delhi 110 002

© 2015 EDITOR
ISBN 9789351305439

Published by : **Daya Publishing House®**
A Division of
Astral International Pvt. Ltd.
– ISO 9001:2008 Certified Company –
House No. 96, Gali No. 6,
Block-C, 30ft Road, Tomar Colony, Burari
New Delhi-110 084
E-mail: info@astralint.com
Website: www.astralint.com

Sales Office : 4760-61/23, Ansari Road, Darya Ganj
New Delhi-110 002   Ph. 011-23245578, 23244987

Laser Typesetting : **Rajender Vashist**
Delhi - 110 059

Printed at : **Replika Press Pvt. Ltd.**

PRINTED IN INDIA

Acknowledgements

It is my privilege and pleasure to express my profound gratitude to our Chief Guest Kun. Vikram Singh, MLA, Rajnagar and Chief Speaker Dr. Nandita Pathak, Director, Deendayal Research Institute Chitrakoot Madhya Pradesh who have very kindly consented for Inaugural Programme of National Conference.

I am heartily thankful to Chief Guest and Chief Speaker Honorable Maharaja Pushpraj Singh Ju Dev, former Education Minister, Govt. of M.P., Rewa, Maharaj, Rewa MP who have very kindly consented for Valedictory Function and Award Ceremony of this National Conference.

I am Thankful to the National Academy of Sciences India, Allahabad for its in association with Environment and Social Welfare Society, Khajuraho for organizing this Conference.

I am highly thankful to Madhya Pradesh Council of Science and Technology, Bhopal for grant to Environment and Social Welfare Society, Khajuraho for this National Conference.

I am thankful to Honorable Guest Dr. Niraj Kumar, Executive Secretary, The National Academy of Sciences India, Allahabad, Dr. M. S. Parihar, Professor and Head, School of Studies in Zoology, Vikram University, Ujjain MP., Dr. Masood Akhtar, Collector and District Magistrate, Chhatarpur. Dr S. K. Mandal, Chief Conservator of Forest, Range Chhatarpur, Mr. Sushil barman, District Co-ordinator, JAP, Chhatarpur for his valuable support and inspiration.

I am heartily thankful to honorable Invitee Guest Dr. Srinivas Murthi,

Chief Conservator of Forest, Panna Tiger Reserve, Panna. Prof. U. C. Srivastava, Department of Zoology, University of Allahabad, Allahabad UP. Dr. Shives Pratap Singh, Professor and Head, Department of Zoology, Government College, Satna MP. Dr. Harish Vyas, Professor and Head, Department of Botany, Govt. Girls Kalidas College, Budhwaria, Ujjain MP. Dr. Rajesh Saxena, Sr. Scientist, Madhya Pradesh Council of Science and Technology, Bhopal MP. Dr. Alka Vyas, Professor of Microbiology, School of Studies in Microbiology, Vikram University, Ujjain MP. Dr. Devendra N. Pandey, Professor of Zoology, Govt. College Maugang, Rewa, Awadhesh Pratap Singh, Rewa. Dr. Veena Pande, Professor of Biotechnology, Kumaun University, Bhimtal, Uttarakhand. Dr. T. R. Nayak, Principal, Government Chhatrasal College, Panna. Dr. K.P. Singh, Professor of Zoology, University of Allahabad, Allahabad UP. Who have very kindly consented and given us an opportunity to share your valuable thought which will provide milestone on the way of leading Scientists in this National Conference.

I am especially thankful to all delegates who actively participated in this National Conference.

I am profoundly thankful to my Board of Director and All members of Environment and Social Welfare Society, Khajuraho for their invaluable cooperation, and all of those person who are directly or indirectly concerned with this conference.

Dr. Ashwani Kumar Dubey
President
National Conference
& Environment and Social Welfare Society,
Khajuraho-471606

Preface

I am extremely happy to great you al all taking part in the National Conference On "Sustainable Development of Natural Resources and Wildlife Conservation" in world heritage place, Khajuraho organized by Environment and Social Welfare Society, Khajuraho In Association with National Academy of Sciences India, Allahabad and Sponsored by Madhya Pradesh Council of Science & Technology, Bhopal. Assist by Godavari Academy of Science & Technology, Chhatarpur. Objective of this conference is to provide a platform to researcher and students to disseminate knowledge related to Natural Resources, Wildlife Conservation, Biodiversity, Environment and Technology. To enable leading scientist to share their ideas with young scholars for development of Global Health. With theme taking some positive steps towards improving our Natural Resources, Wildlife Conservation, Biodiversity, Environment and Technology to protect our Environment and Planet for future generation.

The science-based technology has offered the promise of a better world by the best technology improve in standard of Human beings. But, its negative consequences such as enormous exploitation of natural resources, emission of toxic elements, air and water pollution have created conditions for environmental health and have already caused irreversible damage to the biosphere. Despite the ongoing technological revolution, the majority of world population still lives in abject poverty with inadequate food, water, housing and energy facing the problems of Global health and Safe Environment.

The rich diverse Wildlife in India has been used to a great extent to promote Indian tourism at the international level. Every year hoards of tourists visit the wildlife sanctuaries and parks of the country to take a look at the different kinds of animal and bird species.

Every year a seminar is organized on some recent topic of interest, directly related to environment. There have been many environmental changes due to natural causes resulting in anthropogenic or human induced changes. This is because of the already large and continuing growth of human population, also needs of human being are becoming increasingly consumers with resulting increased greater demand of all necessities, of like such as food, water, energy, housing, transportation and the like. This is putting tremendous pressure on the already depleting natural resources and wild life of the earth system.

Two linked issues which are currently of great include world-wide exploitation of natural resources and Wildlife. We know that the causes of degradation of natural resources and Wildlife. By Natural: Earthquake,Volcano and Fire. By Man: Human has introduced a large number of hazardous substances into the environment. Sewage, Fertilizer, Toxic Chemicals, Oils, Pesticide, Sulpherdioxide, Carbon monoxide, Deforestation, Population explosion, Industrial development, Technical development, Indiscriminate use of resources, Intensive agriculture practices, Faulty use of agronomic, procedures, Overgrazing, Urbanization, Rapid development mean of transport, Pollution, Natural hazards, Cultural problem, Hunting, Destruction of habitat and Over exploitation.

Obviously some chemicals are useful but many are hazardous and their harm to Wild life and environment and our health far outweighs their benefit to society. Therefore, we need to discuss risks factor for safe environment and wildlife.

Now a day's everybody knows that the causes of degradation of Nature, However now the time to take action for Environment conservation. Definitely by this National Conference researcher, scientist, subject expert, student, participant as well as listener will improve his mental IQ and will start to conserve our living kingdom and wildlife.

Dr. Ashwani Kumar Dubey
President
National Conference
& Environment and Social Welfare Society,
Khajuraho-471606

List of Contributors

A.P. Kanungo, Extension Education Department, OUAT, Bhubaneswar, Odisha, India.

Anama Charan Behera, Department of Economics, D.B. College, Turumunga, Keonjhar, Odisha, India.

Anand Kumar Mishra, Department of Zoology, Govt. S.M.S.P.G. College, Shivpuri-473551, M.P., India.

Anil Pandey, P.G. Department of Microbiology, Govt. Madhav Vigyan Mahavidhyala, Ujjain.

Arpita Awasthi, Department of Botany and Microbiology, T.R.S. College, Rewa (M.P.), India.

Ashwani Kumar Dubey, Research & Development Unit, Godavari Academy of Science and Technology, Environment & Social Welfare Society, Chhatarpur-471001, India.

B. Parasar, Extension Education Department, OUAT, Bhubaneswar, Odisha, India.

B.K. Mohanty, Extension Education Department, OUAT, Bhubaneswar, Odisha, India.

B.P. Mohapatra, Extension Education Department, OUAT, Bhubaneswar, Odisha, India.

Bharatendu Prakash, Vikram Sarabhai, Fellow (MPCST-2008-2012), Bundelkhand Resources' Study Centre, Chhatarpur 471001 (M.P.), India.

Bibhu Santosh Behera, College of Agriculture, Department of Extension Education (OUAT), Bhubaneswar, Odisha, India

Deepak Mishra, Department of Biotechnology, A.K.S. University Satna (M.P.), India.

Deepak Vyas, Lab of Microbial Technology and Plant Pathology, Department of Botany, Dr HS Gour University, Sagar (MP), India.

Mangla Bhide, Assistant Professor, Department of Zoology, Dr. H.S. Gour (Central) University, Sagar. (M.P.), India.

Manish K. Sharma, P.G. Department of Microbiology, Govt. Madhav Vigyan Mahavidhyala, Ujjain, India.

Mohit Arya, Department of Zoology, Govt.K.R.G. (Auto.) P.G. College, Gwalior-474004, M.P., India.

Nandita Pathak, Director,J.P. Foundation, Deendayal Research Institute, Chitrakoot District, Satna (MP), India.

Payal Mahobiya, Assistant Professor, Department of Zoology, Dr. H.S. Gour (Central) University, Sagar. (M.P.), India.

Poonam Dehariya, Lab of Microbial Technology and Plant Pathology, Department of Botany, Dr HS Gour University, Sagar (MP), India.

Prahlad Dube, Biodiversity Research Lab, Department of Zoology, Government College, Kota (Rajasthan) India

Pratima Mishra, Department of Biotechnology, Pentium Point College Rewa (M.P.), India.

Praveesh Bhati, P.G. Department of Microbiology, Govt. Madhav Vigyan Mahavidhyala, Ujjain, India.

R.K. Sharma, Department of Mathematics, Government College, Bundi, (Rajasthan), India.

Raj Karan, Research Scholar, Dept. of Energy & Environment, MGCGV, Chitrakoot, Satna-485331, M.P., India.

Rekha Rai, Govt. Girls P.G. Excellence collage Sagar (M.P.), India.

S.K.Rout, Extension Education Department, OUAT, Bhubaneswar, Odisha, India.

Sadhana Chaurasia, Head, Dept. of Energy & Environment, MGCGV, Chitrakoot, Satna-485331, M.P., India.

Seema Sharma, Department of Bioscience & Biotechnology, Banasthali Vidyapeeth (Rajasthan), India.

Usha Pancholi, Department of Mathematics, Government College, Kota-324009, (Rajasthan), India.

Veena Pande, Department of Biotechnology, Bhimtal campus, Kumaun University, Nainital, 263001, India.

नवनीत कुमार गुप्ता, विज्ञान प्रसार, विज्ञान एवं प्रौद्योगिकी विभाग, सी-24, कुतूब संस्थानिक क्षेत्र, नई दिल्ली-16

Contents

1

Sustainable Development of Natural Resources and Wildlife Conservation

Nandita Pathak

ABSTRACT

Natural resources are considered as the most important driver of sustainable economic growth of any Nation as they create jobs and provide governments with revenue to deliver services to their citizens. In a developing Country like India, higher economic growth is necessary to reduce the level of poverty and improve the living standard of ever increasing populace. Higher economic growth will require greater use of natural resources and environment which in turn will lead to their degradation and eventual decay. The pressure on these natural resources is further increasing with increased population. Hence, higher growth may not necessarily lead to sustainable development unless it is accompanied by environmental protection. Thus, it is a big challenge before us to manage and use our natural resources in a sustainable manner.

Introduction

Our country being a home to rich and diverse wildlife flora and fauna because of wide range of climate, soil, weather and other such factors. Owing to such diversity, equal number of rare as well as threatened animals and plants are found that need to be protected. This calls for much greater wildlife conservation efforts in India.

Nandita Pathak, J.P. Foundation, Deendayal Research Institute, Chitrakoot District, Satna (MP).

Land and water are the two most precious natural resources of any nation. They are vital to our very existence. It is, therefore, imperative that we manage and conserve them to meet the growing need for food, fodder, fiber and fuel. In future, our abilityto meet the growing demand of the increasing population will depend on how efficiently we use these resources without reducing their potential. Because of the multiplicity of uses, the pressure on the available natural resources is increasing, and gradually, less and less resources will be available for meeting the requirements of food, feed and fodder of the exploding human and livestock population. As a result of increasing population, which doubled every 25 to 30 years during the 20th century, pressure on these resources had increased, and degradation of these resources has become a worldwide concern. And now it is being increasingly felt that land degradation and water shortage/ pollution effects and impacts represent a serious threat not only to the environment, but also to the survival of millions of people living in such areas. At the same time, while national economic growth has shown a steady increase over the years up to some extent, the poverty in several parts of the country especially in rural areas is still a big concern. This situation stems from the mismanagement of land and water resources.

According to the Report of the National Remote Sensing Agency in Hyderabad, the country has 63.85 million hectares of wastelands, which account for 20.17% of the total geographical area (328.72 million hectares) and in Madhya Pradesh, 59.34 lakh hectares (19.3 %) of the total geographical area of 307.44 lakh are wastelands. The large percentage of which is under pastures and grazing lands, gullied/ravenous land, degraded notified forestland, and barren rocky areas. The situation has not improved since. Increased population pressures and a lack of forest management, had also led to a gradual degradation of forest lands in many parts of the country, where forests and trees provide rural people not only wood but a range of other products and services. As a substitute to these vanishing resources, economic tree planting has become, increasingly, a pronounced need in rehabilitating such degraded forest lands. It is roughly estimated that, at the least, more than half of the forest lands are degraded. Of late, there has been a realization that immediate measures are required to rehabilitate degraded soils that have been rendered wastelands due to partial or complete loss in productivity.

India has vast tracts of wastelands, which have been lying barren for ages. Most of such lands are suitable for growing certain economic trees and thus could be put to socially productive uses. Although a number of initiatives have been taken by the central and some state governments, for turning these wastelands into cultivable land, the rate of conversion has been quite slow due to a lack of suitable technologies. A major constraint on the pace of expansion has been the non-availability of suitable technologies for the rehabilitation of degraded land.

Our experience in Chitrakoot area is that the development and management of these two natural resources particularly land and water will not only improve the health of the soil and environment, but will also help improve the nutritional levels of the Indian populace if fruit orchards are developed on such lands, so important to make the nation healthy in all respects. In a country like India where malnutrition is a big challenge before the governments, As such, fruit culture on wastelands and degraded forest lands became of major importance in India, due to increasing needs for fruits, and the realization that certain fruits could be grown successfully on degraded wastelands with the adoption of suitable technology. So far, there is relatively little experience in the rehabilitation process of such degraded soils through fruit plantation, and research is urgently needed on the standardization of suitable rehabilitation technologies. What is needed now is to strengthen on-going efforts at the national as well as regional (or ecological) level, providing not only nutritional security but also ensuring economic security.

Wasteland development through fruit culture would help to attain twin national goals, reducing poverty and increasing income & employment opportunities in rural areas. Since wastelands/panchayat lands are spread all over the country, they provide countrywide opportunities for becoming economic hubs. Since fruit plant cultivation is a labour-oriented activity, they can become employment generation hubs. Thus, rehabilitation of such degraded lands through fruit plants will not only help in reducing the drudgery and generating gainful employment and self-employment opportunities, but will also help in conservation of soil and scarce water resources.

Conclusion

Keeping in view the importance of natural resources in the economic growth of our country, there is an urgent need to give immediate attention towards conservation and rehabilitation of these resources. The systematic efforts are required towards the standardization of rehabilitation technology.

At the same time, Wildlife conservation protecting endangered plant and animal species and their habitats to ensure that nature will be around for future generations has become an increasingly important practice due to the negative effects of human activity on wildlife. India is considered a home to rich and diverse wildlife flora and fauna because of wide range of climate, soil, weather and other such factors. Owing to such diversity, equal number of rare as well as threatened animals and plants are found that need to be protected. This leads to the need of much greater wildlife conservation efforts in India. As per the survey India is a home to about 60-70% of the total biodiversity found across the world and about 33% of plant species are endemic. There are about 2.9% threatened species in India. This further enforces the need of right wildlife conservation efforts in India. Due to the growing impact of deforestation, continuous efforts are needed to protect the endangered species of wildlife as well as those that are on the verge of extinction.

Sustainable Development of Natural Resources and   *Pages* **5-17**
Wildlife Conservation
Editor: **Ashwani Kumar Dubey**
Published by: **REGENCY PUBLICATIONS, NEW DELHI**

2

Natural Resources Need Sustainable Management

Bharatendu Prakash

If we review the past 66 years of scenario in India, the policies as well as approach adopted during the post-colonial period did not differ much from the earlier way of thinking or working during colonial era. In spite of democratic pattern of governance the mind set remained the same. Forests which were confiscated by British Government in the year around 1865 could not be restored to villages and tribal communities. The destruction of forests and wildlife and neglect of traditional and natural resources which had started under foreign dominance multiplied several times during post-independence era. The development policies which were adopted never gave any importance to local-regional culture, climate and potentials. For our leaders urbanization, industrialization and commercialization and later globalization became synonyms of development . Thus over-exploitation of forests, hills, minerals, water resources, ground water reserve and soils went on without any check and sense of balance.

Above situation resulted into some very serious problems related to environment, available resources and human life. The crisis with this sort of development based on exploitation of resources during all these decades has been felt all over the world. The terms like Sustainable Development, Natural Resources Conservation, Renewable Energy, Bio-diversity Protection and Sustainable Agriculture etc. came into popular use in this era only. We hear so much about global warming and the climate change which are

Vikram Sarabhai, Fellow (MPCST-2008-2012), Bundelkhand Resources' Study Centre Chhatarpur 471001 (M.P.)

essentially outcome of the problems created by human being's mishandling of entire nature. Despite every society facing all such problems transformation from old pattern towards better alternatives is extremely slow and it is more so with passive societies like India.

Some of the key issues mentioned above are being presented below in slight detail for greater understanding with respect to what is being thought of at global scale.

SUSTAINABLE DEVELOPMENT

A general definition of Sustainable Development can be taken as a mode of human development in which resource-use aims to meet human needs while ensuring the sustainability of natural systems and the environment, in a way that these needs can be met not only in the present, but also for generations to come.

The term 'sustainable development' was first used by the Brundtland Commission (1987) which coined what has become the most often-quoted definition of sustainable development: "development that meets the needs of the present without compromising the ability of future generations to meet their own needs."

The concept of sustainable development has in the past most often been broken out into three constituents: environmental sustainability, economic sustainability and socio-political sustainability . More recently another important constituent, the cultural sustainability has been added to above three domains.

According to Brundtland Commission's report the above definition contained within it two key concepts:

- The concept of 'needs', in particular the essential needs of the world's poor, to which overriding priority should be given; and
- The idea of limitations imposed by the state of technology and social organization on the environment's ability to meet present and future needs.

A recent UN- document has discussed its various aspects including defining four major dimensions of sustainable development :

A. economic development

B. social inclusion

C. environmental sustainability

D. good governance along with four related normative concerns *viz.*

- The right to development,
- Human rights and social inclusion
- Convergence of living standards, and
- Shared responsibilities and opportunities.

In the same line of thought the Leadership Council of UN-Sustainable Development Solutions Network (SDSN) identified following ten challenges:

1. ending extreme poverty including hunger,
2. achieving development within planetary boundaries.
3. ensure effective learning for all children and youth for life and livelihood,
4. achieving gender equality, social inclusion and human rights for all,
5. achieving health and well being at all ages,
6. improving agriculture systems and raise rural prosperity,
7. empowering inclusive, productive and resilient cities,
8. curbing human induced climate change and ensuring sustainable energy,
9. securing ecosystem services and bio-diversity and ensure good management, and
10. transforming governance for sustainable development.

UN-direction for these ten sustainable development challenges has been " need to be addressed at global, regional, national and local scales." It desired that at each level a plausible basis for framing the Sustainable Development Goals (SDGs) could be framed to trigger practical solutions that governments, businessmen, and civil society can pursue with high priority. Particular emphasis is to be given so that children everywhere learn the SDGs to help them understand the challenges that they will confront later as adults.

Aforesaid Sustainable Development- Challenges are inherently integrated, thus it was desirable that these need to be pursued in combination, rather than individually or one at a time.

Let us understand various aspects of Sustainable Development:

Environmental Sustainability is the process of making sure that current processes of interaction with the environment are pursued with idea of keeping the environment as pristine as naturally possible. Thus, environmental sustainability demands that society designs activities to meet human needs while indefinitely preserving the life support systems of the planet. This requires sustainable use of water, utilization of only the renewable energy and maintain sustainable material supplies (e.g. harvesting wood from forests at a rate that maintains the biomass and biodiversity).

An "unsustainable situation" occurs when the natural capital (the sum total of nature's resources) is used up faster than it can be replenished. Sustainability requires that human activity uses nature's resources only at a rate at which these can be replenished naturally. Inherently the concept of sustainable development is intertwined with the concept of carrying capacity. Theoretically, the long-term result of environmental degradation is the inability to sustain human life. Such degradation on a global scale should imply extinction for humanity.

CONSERVATION OF NATURAL RESOURCES

The earth has plenty of resources that people have been using to meet their needs. The earth is a great place where both living and non living things can be found. The living things include plants and animals while the non-living things are the land, water and air . The materials on earth that people use are called natural resources. There are two types of natural resources on the earth; Renewable and Nonrenewable both discussed below:

RENEWABLE RESOURCES

Human beings need both plants and animals for their survival. All the food people eat come from plants and animals. The plants (called crops) which people eat are replaced by new ones after each harvest. Some people also eat animals. Animals have the capacity to reproduce and are replaced as the young animals are born. Plants and animals are resources that can be replaced. Water in a river or in a well may dry up but when rain comes the water is replaced. Water is a resource that can be replaced. They are called renewable resources.

NON-RENEWABLE RESOURCES

Most plants grow in top soil. What happens to soil when there are rains and floods? Rains and floods wash away the top soils. Soil is formed from rocks and materials from dead plants and animals. It takes thousand of years for soil to get formed. Therefore, it cannot be replaced easily. It is a nonrenewable resource.

There are rich supplies of iron and aluminum in the earth. But people are using them up fast. They have to dig deeper into the earth to get what they need. Coal, oil and natural gas are called fossil fuels. These were formed from plants and animals that lived on earth millions of years ago and buried deep in the earth. In present form these are harvested by man through oil rigs and mining. These fuels are used to produce electricity, machine fuels and other things to meet people's needs. Presently very huge amount of fossil fuels are being used up. Scientists say that there will come a time when these fuels will be finished. As mentioned earlier it takes millions of years for dead plants and animals to turn into fossil fuels. Once these fuels are used up there will not be replacement. These are called nonrenewable resources.

DESTRUCTION OF NATURAL RESOURCES BEYOND LIMIT

In present society there is increase in activities which are harmful to both existing renewable and nonrenewable resources. To build new roads and multi-story RCC buildings, mountains are blasted off killing both plants and animals. It also creates destruction to their natural habitat. Rice fields and water reservoirs are turned into residential or commercial centers. Too many trees are felled each day and sold for easy profits without thinking that when all the trees are gone, rains will wash away the top-soils making earth an un-productive mass. In severe deforested situation even rainfall becomes irregular and uncertain .

With emphasis on industrialization and urbanization, mankind's great demand for natural resources and its large scale exploitation and consumption has resulted in the weakening, deterioration and exhaustion of these resources. One difficult task faced presently is to guarantee the lasting utilization of natural resources at the lowest possible environmental cost while still assuring economic and social development. The traditional mode of resource consuming development and the current inefficient economy are seriously threatening the lasting utilization of natural resources

Industrialization brings people to build factories. If not properly planned and maintained, these factories emit harmful waste materials that pollute the soil, air and water. Too much mining and quarrying for the purpose of getting precious metals and stones eventually destroys the forests, hills and erodes the soil. Some farmers use too much chemical fertilizers which destroy the quality of the soil and are harmful to both human and animals. It has been observed that even normal application of chemicals to soil damages the natural micro-organism responsible for soil-rebuilding at the farm. Burning plastics and garbage, and smoke belching cars also pollute the air and contribute to global warming.

All this is forcing people to give maximum priority to conservation of resources. There are many ways that one can conserve natural resources. One has to look around and see what natural resources are being used and think the way to limit the usage or choose alternatives affecting the environment least. Natural resource like trees could be conserved through more plantation and recycling process. Many products come from the trees like papers, cups, cardboards and envelopes. By recycling these products one can reduce the number of trees cut down a year. One should make the most use of these paper products without being wasteful and then recycle them. Recycling is an important activity for conserving natural resources.

RENEWABLE ENERGY

Major energy usage today is of coal for thermal power plants, other fossil fuels like petrol, diesel, crude oil and the natural gas in day-to-day life in transportation, traction, running pumps, cooking, manufacture of fertilizers, preparation of drugs, moving railway-trains and aero-planes etc. The discovery of fossil fuel had added to life greater comfort and control over nature in many ways. This naturally upsets many people when conservation is talked about.

Finding ways for conserving natural resources and seeking alternatives has come as a challenge for common man, industrialists, governments and the scientists. In energy sector options are Hydro-power, Solar- and Wind-power. Power can be generated from these sources, and that is the best way for conservation of natural resources such as the fossil fuels. In fact these alternate energy sources are clean and healthy for environment also. Moreover, these energy sources do not emit or produce harmful gases or toxin into our environment like that of the burning fossil fuels. These are renewable as well as difficult to be easily depleted.

BIODIVERSITY - ITS PROTECTION

Nature has been so generous that it gave enormous bio-diversity on earth seen as innumerable flora and fauna. Human beings are also a part of that huge plan of nature. The law of nature includes inter-dependence of all life forms which are complimentary to each other. Thus plants, animals, and other creatures including invisible micro-organism all have an important role to play.

As was discussed earlier the trend of development which considers urbanization and industrialization as major goals has upset entire environment, larger share of which is bio-diversity. Today the impact of global warming and climate change has tried to open many eyes towards this severe problem faced by humanity against its own survival.

Biodiversity in fact helps to balance the nature and also has its own economic importance. Protecting biodiversity has come to be the most important task since biodiversity plays crucial role in reducing pollution, dealing as well as adapting to the effects of climate change. If we don't act immediately the effects could be as harmful as the effects of global warming itself. The tropical forests which are critical to fighting climate change and home to more species than any other ecosystem type, have to be taken care of very urgently.

We know that farm crops as well as animals are mainly descendents of wild organisms and these are generally important component of biodiversity. Some varieties of old crops have more taste or disease resistance, and these may be more suitable to future changes in the climate. Most of the fruit crops depend on various insects for pollination of their flowers. Biodiversity is essential also as provider of natural services, though we may not readily recognize it. For instance Peat bogs play an important role in purifying water and also lock up carbon dioxide Tiny plants that grow also absorb large amount of carbon dioxide from the atmosphere.

Biodiversity is an important part of sustainable development which should be considered as a major target for industry as well as planning system because monitoring and measuring biodiversity is the only way to achieve this target.

To be precise biodiversity protection is extremely important since biodiversity is a fundamental component of life on Earth. It creates complex ecosystems that could never be reproduced by humans. The value of biodiversity, both intrinsically and to humans, is immeasurable, and thus must be protected.

SUSTAINABLE AGRICULTURE

Sustainable Agriculture may be defined as consisting of environmentally friendly methods of crops or livestock without damage to human or natural systems. More specifically, it might be said to include preventing adverse effects to soil, water, biodiversity, surrounding or downstream resources— as well as to those working or living on the farm or in neighboring areas. Furthermore, the concept of sustainable agriculture extends inter-generationally, relating to passing on a conserved or improved natural resource, biotic, and economic base instead of one which has been depleted or polluted. Sustainable Agriculture in popular language can be understood as Natural Organic Agriculture which comprises some of the following forms :

MIXED FARMING

Mixed farming is a process of managing a mixture of different crops or animals. The best known form of mixing occurs probably where crop residues are used to feed the animals and the excreta from animals are used as nutrients for the crop. Such farming has been popularly and traditionally followed by all Indian farmers up to a few decades ago until Green Revolution techniques did plunder.

Other forms of mixing takes place where grazing under fruit trees keeps the grass short or where manure from pigs is used to feed the fish. Mixed farming exists in many forms depending on external and internal factors. External factors are: weather patterns, market prices, political stability and technological development while Internal factors relate to local soil characteristics, composition of family and farmer's ingenuity.

Mixed farming provides farmers with :

(a) an opportunity to diversify risk from single-crop production;

(b) use of labour more efficiently;

(c) having a source of cash for purchasing farm inputs;

(d) adding value to crop or crop by-product;

(e) combining crops and livestock.

MULTIPLE CROPPING

Multiple cropping is the process of growing two or more crops in the same piece of land, during the same season. It can be rightly called a form of poly-culture which includes:

(a) Double Cropping (the practice where second crop is planted after the first has been harvested).

(b) Relay Cropping (the practice where a second crop is started along with the first one, before it is harvested).

CROP ROTATION

Crop-rotation is process of growing two or more dissimilar or unrelated crops in the same piece of land in different seasons . This process could be adopted as it has important benefits listed as follows:

(a) this avoids building up of pests that often occur when one species is continuously cropped,

(b) replenishes nitrogen through the use of green manure in sequence with cereals and other crops ;

(c) improves soil structure and fertility by alternating deep-rooted and shallow-rooted plants.

AGRO-FORESTRY

Agro-Forestry is a collective name for land use systems and practices in which woody perennials are deliberately integrated with crops and/or animals on the same land management unit. The integration can be either in a spatial mixture or in a temporal sequence. There are normally both ecological and economic interactions between woody and non-woody components in agro-forestry.

ROLE OF SCIENCE AND TECHNOLOGY

Since present era is dominated by Science & Technology (S&T), it seems desirable that the role of science & technology be explored in the process of sustainable development of a nation or region.

S&T can play very vital role in helping society to move towards sustainable development which has to be a scientific and technological endeavor that seeks to enhance the contribution of knowledge to environmentally sustainable human development around the world This emerging enterprise is focused on deepening our understanding of socio-ecological systems in particular places while exploring innovative mechanisms for producing knowledge so that it is relevant, credible, and legitimate to local decision makers.

REORIENTING TECHNOLOGY

Fulfilling aforesaid task requires reorientation of technology the key link between humans and nature. The development in this direction must be changed to pay greater attention to environmental factors.

In countries like India capacity for technological innovation needs to be greatly enhanced in order to get proper and effective response towards the challenges of sustainable development. The technologies of industrial countries are not always suited or easily adaptable to the socio-economic and environmental conditions of such developing countries. To compound the problem, the bulk of world research and development (R & D) addresses few of the pressing issues facing these countries, such as arid-land agriculture or the control of tropical diseases etc.. Not enough is being done to adapt recent innovations in materials technology, energy conservation, information technology, and biotechnology to the needs of developing countries.

These gaps must be covered by enhancing research, design, development, and extension capabilities in the developing countries.

RESPONSIBILITY OF TECHNOLOGY

the processes of generating alternative technologies, upgrading traditional ones, and selecting and adapting imported technologies should be governed by environmental resource concerns. Most technological research by commercial organizations is devoted to product and process innovations that have market value. Technologies are needed that produce 'social goods', such as improved air quality or increased product life, or that resolve problems normally outside the cost calculus of individual enterprises, such as the external costs of pollution or waste disposal.

SPECIAL ISSUES AFFECTING BUNDELKHAND

After discussing major environmental issues as it relates globally and the role of technology, major problems which have been expressed by people of *Bundelkhand* need to be recapitulated in order to explore possible technology and intervention alternatives to move this region towards sustainable development. These problems include broadly:

- Neglect of traditional knowledge, culture, resources, seeds and essential needs,
- Destruction of forests, hills, traditional water bodies and damming/ diverting the rivers,
- Non-committed and loose functioning of Govt. schemes serving rural areas,
- Emphasis over cost intensive commercial farming as against cost effective integrated organic farming,

- Starving villages and pushing forward the urbanization & industrialization agenda,
- Institutions not attending local needs of technologies and appropriate education,
- No effort to encourage small enterprises locally to curtail forced migration,
- Women & children facing malnourishment and being neglected .
- Villages facing poor infra-structure, roads, schools, health centres, veterinary services, water-shed development, ware-houses, communication and electricity etc.
- Feudal atmosphere still persisting which suppresses poor and downtrodden people.

Tackling above problems needs to be pursued at three major levels: Government, Industry and People, which presently are either unconcerned or incapable to take proper action discussed briefly as follows:

1. THE GOVERNMENT LEVEL

The references which we get from United Nations ' documents clearly devise strong and immediate measures to be taken at government level. Further planning of any development project, adopting industrial policies, promoting education & research as well as exploitation of natural resources should be governed by measures ensuring sustainable development of nation as well as the region.

Unfortunately our governments run the show by passing ordinances without placing many plans and proposals even before Parliament for discussion and approval. What suits the international pressure and industrial houses becomes rule here. Thus at the level of governance short sighted plans get pursued under anti-nature & anti-people policies which endanger the environment much more leading people to face enormous difficulties and ultimately the human race towards struggling for its own survival. UN in all these situations mostly remains a silent observer.

2. THE INDUSTRY LEVEL

There is hardly any industry in India which cares for environment the least. Governments view industrialization as progress of the country and promote everything which can raise its revenue and profits to the industries. Whenever some voice is raised they quote certain arithmetic figures (percentages) in projecting continuous development irrespective of the deep crisis and increasing worries of common people. In regions like *Bundelkhand* what is being promoted and pursued in name of industry is mainly exploiting the natural resources beyond limit. Large number of mining and stone crushing plants are proliferating and a simple calculation of granite crushing leads to a figure of 50,000 metric- tonne of stone chips being transported out every day. The extraction of diamond from Panna and the new ground opened at Buxwaha block of Chhatarpur is leading irreversible damage to the forests, fertile top-soils, surface water storage and ground water recharge.

In this endeavor central government and the state government seem to have identical views. Destruction of forests, damage to Panna Tiger Reserve (spread in Panna and Chhatarpur districts), drying up of the perennial *Ken*, *Dhasan* and *Betwa* rivers, reducing annual rain-fall and farmers committing suicides have no impact on governmental planning & working. Whether it is promotion of industries, adoption of chemical intensive mechanized commercial farming or planning for new power projects there is no concern of environment which may direct them to take appropriate action adopting environment-friendly approach.

3. THE PEOPLES' LEVEL

There are many individuals, people's groups and environmentally active voluntary (civil society) organizations who raise issues demanding immediate action to abandon the plans, projects, and industries which damage environment. There have been several moves for protecting forests, stop mining, cancelling nuclear power plants, reducing dam-heights and saving important rivers *Ganga*, *Yamuna* and *Narmada* along with other smaller ones, but each time such voices are brutally suppressed by governments and the all powerful industrial corporations . It is surprising that the United Nations again maintains silence although theoretically this body is said to believe in sustainable development process.

Rich farmers' organizations and the industrial houses do not care and continue pursue their own agenda which is unfortunately backed by the governments. A few important Public Interest Litigations (PIL) on environmental issues have been filed before country's highest court. Some favourable orders have also been passed but implementation of such orders is mostly delayed purposely. In this way any little resistance which people try to build up on certain pressing issues just fades out and one such shock is enough to silence people for several years to come.

MOVING TOWARDS STRATEGY-ALTERNATIVES

In the light of above situation which draws a dismal picture of present day organized efforts towards developing India or the regions like *Bundelkhand* with an environmental concern minimizing people's miseries and save this region from turning into irreparable deep cratered dry desert, the only possibility lies with the institutions with sensitive & concerned scientists working with an aim to pursue well- being of common people and the conscious individuals of society who are committed to raise people's conscious awareness regarding present status of dwindling resources and the measures needed to adopt alternative path of sustainable development.

DIRECTION FOR CONCENTRATED EFFORTS

In any area of work success is positively met if one starts building up on the basis of inherent potentials. That inspires us to first see what could be done in this region based on the potentials of *Bundelkhand*. Recapitulating potentials related to the natural and created resources of *Bundelkhand* we come across following areas of strength:

a. Background of Rich Forests which could be regenerated with proper efforts,

b. Natural Water Resources: Rivers and Rivulets along with Created Reservoirs,

c. People's inherent water-conserving culture,

d. Varieties of soils capable of growing numerous coarse-grains and normal crops organically,

e. Indigenous Creative Knowledge accumulated in course of past centuries.

These potentials are sufficient to initiate a development process which ensures protection of environment, sustenance of natural resources as well as creation of rural based small productive enterprises to turn the road of common people's migration back towards their villages / homes.

INSTITUTIONS TO CARRY FORWARD SUSTAINABLE ALTRERNATIVES

Essential steps to be taken up for initiating sustainable development process in such regions will fall ultimately on the shoulders of following institutions:

1. Existing Science-Technology Institutions such as State S-T councils, DST, Indian Institute of Technology's, Indian Institute of Science, and NIT's etc.

2. People controlled Educational Centres,

3. Self-sustaining People's Organizations and Voluntary groups,

4. Nagar-Panchayats and Gram-Panchayats,

5. Appropriate new Initiatives like:

    • Appropriate Technology Resource Centres,

    • The Science-Academies including panels and associations of people friendly and environmentally sensitive Scientists,

    • Training and Extension Institutes set up in rural areas preferably at Block-level,

    • People's Marketing Associations active from village-level up to cities to coordinate marketing of organic and natural products from villages. .

Fortunately this country has several such prestigious S & T Institutions, Independent Colleges, Private Universities, Science Academies, vast network of Gram-Panchayats and Nagar Panchayats headed by people's representatives, Farmers' and Peoples' organizations and networks like OFAI, ASHA and BKU etc. who could be involved in research, development, extension, trainings and all necessary actions to work towards environment friendly processes and projects.

SOME BROAD AREAS WHERE S.& T. INTERVENTION CAN BE HELPFUL

In *Bundelkhand* region following areas could be sorted where S&T intervention can bear fruits and help people to overcome several of their problems without affecting the environment and exploiting exhaustible natural resources .

1. Utilization of fast flowing water in seasonal rivulets and perennial rivers for decentralized power-generation making thermal or nuclear options unnecessary,
2. Utilization of solar radiation in generation of electricity and lighting emergency lamps, especially in rural areas,
3. Utilization of seasonal wind –power for generation of low-cost power,
4. Utilization of waste biomass and farm residues along with the cattle-dung and urine for enriching soils,
5. Strengthening animal husbandry giving preference to cows,
6. Development of small agriculture-tools and bullock operated devices,
7. Low input and local resource based farming,
8. Multiplication of indigenous seeds,
9. Regeneration of endangered plant-species and herbs,
10. Preservation of available fruits and vegetables,
11. Value-addition through processing of indigenous cereals, pulses and oil-seeds organically grown in this region (the whole world is our market),
12. Utilization of available over 400 herbs in making medicinal preparations to meet all kinds of diseases,
13. Field- and home—harvesting of rain water for irrigation as well as other purposes,
14. Innovating building / road technologies to minimize destruction of hills/ forests,
15. Improving traditional rural handicrafts using natural products,
16. Documentation and re-validation of traditional knowledge in present circumstances.

Sustainable Development of Natural Resources and
Wildlife Conservation

Pages **18-30**

Editor: **Ashwani Kumar Dubey**
Published by: **REGENCY PUBLICATIONS, NEW DELHI**

3

A Study of Wetland Birds in Sakhya Sagar and Madhav Lakes, Madhav National Park, Shivpuri, Madhya Pradesh

Mohit Arya[1]* and Anand Kumar Mishra[2]

ABSTRACT

The study was carried out at Sakhya Sagar and Madhav Lakes which is located inside the Madhav National Park, Shivpuri, M.P., India. The present survey of wetland birds were performed to bring out the species richness, various locations and habitats utilized by migratory and resident birds. The present survey revealed presence of 73 species of wetland birds belonging to 18 families and 7 orders. Out of 73 species of wetland birds 25 species were residents, 29 resident migrants and 19 migrants; 38 winter visitors, 23 vagrant and 12 species were seen in every season; 26 were common, 22 fairly common, 20 uncommon and 05 species were rarely seen observed during the survey period. 15 locations and 10 habitats were noted to assess the presence and activities of wetland birds in and around the water bodies.

Key words: *Wetland Birds, Sakhya Sagar and Madhav Lakes, Diversity, Habitat Utilization Distribution.*

1. Department of Zoology, Govt.K.R.G. (Auto.) P.G. College, Gwalior-474004, M.P., India.
2. Department of Zoology, Govt.S.M.S.P.G. College, Shivpuri-473551, M.P., India.
*Corresponding author: Email: aryans.mohit@gmail.com; anand81795@gmail.com

Introduction

India is among the top ten nations of the world for high levels of biodiversity. Its immense biological diversity represents about 7% of the world's flora and 6.5% of its fauna. It embraces 10 biogeographically zones and 26 biotic provinces (Rodgers *et al.*, 2000). In India, exhibit significant ecological diversity, primarily because of the variability of climate conditions and the changing topography (Islam, 2006). The wetland habitats of India are under severe stress as seen at the global level. They are unique in the context of their diversity and are a natural abode for several species of birds. The wetland supports a wide variety of flora and fauna. Diverse flora provide ideal habitat in the form of food and shelter for a large number of avifauna. Due to biotic interaction and natural selection process a characteristic relationship between vegetation and the avifauna has developed. Birds are a familiar feature of not only arboreal environment but aquatic environment too. Every one noticed them due to their varied life styles, conspicuousness diurnal habits, interesting plumage and calls. Birds are also regarded as good subject for exploring a number of question of ecological and conservation significance (Urfi, 2003).

The present study is focused not only on preparing the checklist of wetland birds, but also to find out their occurrence and distribution status as well as to create awareness for their conservation. In addition, the study aims at providing the basic information of the avifauna for further studies related to biodiversity of Madhav National Parks, Shivpuri.

Study Area

Shivpuri town in the state of Madhya Pradesh was once the summer capital and formal hunting reserve of the Maharaja of Gwalior. Madhav National Park, Shivpuri was established in the year 1958 with an area of 165.321 sq. km. Extension area was added in 1982 and 1999. Presently total area of the Park is 354.612 sq. km. Madhav National Park is one of the oldest National Parks in the country and it is unique for its scenic beauty and cultural heritage. The Madhav National Park is one of the 9 National Parks of Madhya Pradesh. Two National Highways *viz.* Agra-Bombay (NH-3) and Jhansi-Shivpuri (NH-25) pass through the park and connect it to the important cities. In order to provide a permanent source of water supply, Maharaja of Gwalior built dams across the river Maniyar in the year 1918 to create a chain of lakes. The Sakhya Sagar and Madhav Lakes are spread in an area of 309.01 hectares and 49 hectares respectively and fall within the area of the Madhav National Park. The Sakhya Sagar Lake lies between 77° 43'E longitude and 25° 26' N latitude and the Madhav Lake lies between 77° 44'E longitude and 25°26' N latitude (Fig. 3.1). These are man-made lakes and about 25 sq. km of forest is spread around the lakes. These lakes are important biodiversity support systems and not only add to the natural beauty of the area, but also provide a permanent source of water to the wildlife, and a fine wetland habitat to the aquatic fauna including thousands of migratory birds.

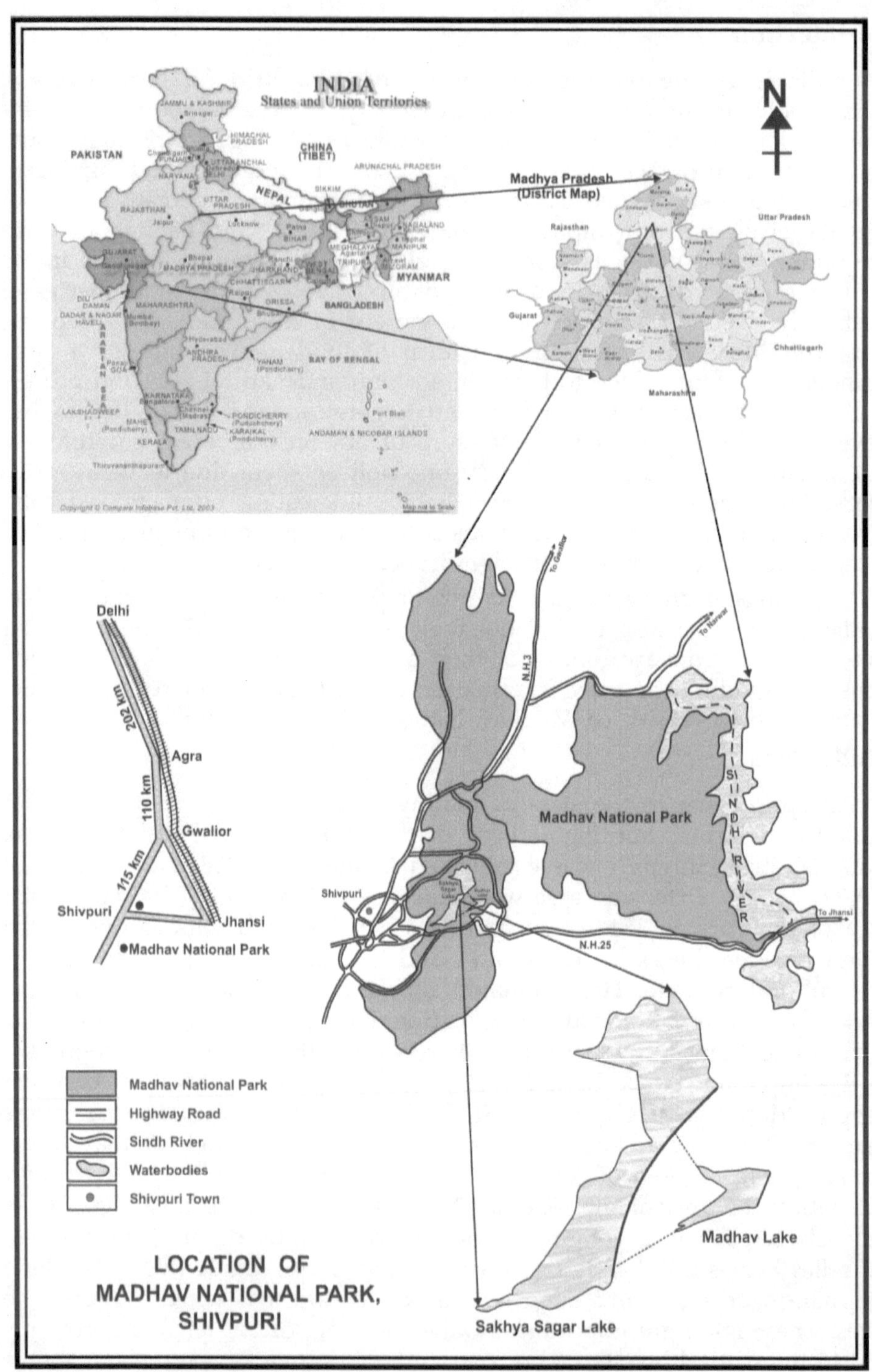

Fig. 3.1. Location map of Madhav National Park, Shivpuri showing wetlands (Sakhya Sagar Lake and Madhav Lake) under study.

Materials and Methods

The study was carried out from January, 2007 to December, 2008. Regular surveys were done in and around Sakhya Sagar and Madhav Lakes to enlist the species of wetland birds throughout the survey period. The survey was conducted from early morning (6:00 to 10:00) and late afternoon (3:00 to 7:00). Direct visual count method was followed for identification and recording number of wetland birds species. Standard guides such as Ali and Ripley (1987), Sonobe and Usui (1993), Grimmet *et al.*, (1999) and Ali (2002) were referred. In an attempt to investigate the wetland birds observed in the study area they were categorized according to their occurrence throughout the study period. The occurrence of wetland birds recorded in the present study was categories as status, seasonal visit and sight frequency. The status of bird species is divided in to three types respectively as Migrants (M), Resident Migrants (RM) and Resident (R). Seasonal visit of these birds is categorized as Winter Visitors (WV), Vagrant

Fig. 3.2. Map of Sakhya Sagar Lake and Madhav Lake showing different location sites.

(V) and Every Season (ES). Similarly the sight frequency is divided in to four types as Common (C), Fairly Common (FC), Uncommon (U) and rarely seen species (Rr). During the present survey the distribution of wetland birds was recorded on the basis of different locations utilized by the wetland birds are as follow; Near Bhadaiya Kund- (BK), Landing No.-1 (L-1), Landing No.-2 (L-2), Landing No.-3 (L-3), Landing No.-4 (L-4), Landing No.-5 (L-5), Landing No.-6 (L-6), Landing No.-7 (L-7), Landing No.-8 (L-8), Landing No.-9 (L-9), Landing No.-10 (L-10), Near Saling Club- (SC), Near Pump House- (PH), Mound Area in Water- (MAW) Northern bank of Lake- (NB). On the basis of utilization of different habitats by wetland birds, 10 habitats were identified as Submerged Vegetation (H-1), Emergent Vegetation (H-2), Floating Vegetation (H-3), Open Water (H-4), Muddy Shoreline (H-5), Shallow marshy area (H-6), Rocky area near water (H-7), Tree and Bushes standing near water (H-8), Reeds and Grasslands near water (H-9) and Fly over and around water bodies (H-10) (Fig. 3.2).

Results and Discussion

The present survey revealed presence of 73 species of wetland birds belonging to 18 families and 7 orders. Species wise percentages of all 18 families are given in table 3.1 and fig. 3.3. The family wise proportion of species richness of wetland birds varied from 24.66% to 1.37%. Family Anatidae with 18 species (24.66%), followed by Charadriidae 10 (13.7%),

Table 3.1. Total number and percentage of wetland bird species of different families

S. No.	Families	Total Species	Percentage
1.	Podicipitidae	1	1.37 %
2.	Pelecanidae	1	1.37 %
3.	Phalacrocoracidae	4	5.47 %
4.	Ardeidae	9	12.33 %
5.	Ciconiidae	6	8.22 %
6.	Threskiornithidae	3	4.11 %
7.	Anatidae	18	24.66 %
8.	Accipitridae	2	2.74 %
9.	Gruidae	2	2.74 %
10.	Rallidae	4	5.48 %
11.	Jacanidae	2	2.74 %
12.	Charadriidae	10	13.70 %
13.	Rostratulidae	1	1.37 %
14.	Recurvirostridae	1	1.37 %
15.	Burhinidae	1	1.37 %
16.	Glareolidae	1	1.37 %
17.	Laridae	4	5.48 %
18.	Alcedinidae	3	4.11 %
Total	**18**	**73**	

Ardeidae 9 (12.33%), Ciconiidae 6 (8.22%), Phalacrocoracidae, Rallidae and Laridae with 4 species each (5.47%), Threskiornithidae and Alcedinidae with 3 species each (4.11%), Accipitridae, Gruidae and Jacanidae with 2 species each (2.74%) and Podiciptidae, Pelecanidae, Rostratulidae, Recurvirostridae, Burhinidae and Glareolidae by a single species each (1.37%). Individuals of family Ardeidae, Podiciptidae, Rallidae, Jacanidae, Rostratulidae, Recurvirostridae, Burhinidae and Alcedinidae were observed to be dominant throughout the survey period.

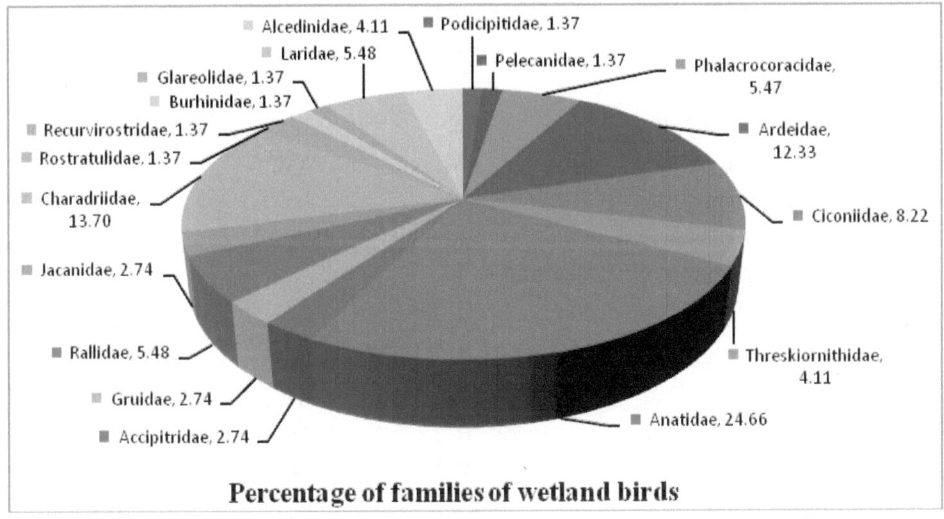

Fig. 3.3. Percentage of different families of wetland bird species in Sakhya Sagar and Madhav Lakes

Out of 73 species of birds found in Sakhya Sagar and Madhav Lakes, 25 species were residents, 29 resident migrants and 19 migrants. 38 winter visitors, 23 vagrant and 12 species were seen in every season. 26 were common, 22 fairly common, 20 uncommon and 05 species were rarely seen (Tables 3.2, 3.3 and Fig. 3.4).

Sakhya Sagar Lake and Madhav Lake were found to be utilizing by different wetland bird species. The distribution of wetland birds is based on availability of food at various locations, where the different bird species were recorded frequently. A total of 12 location sites in Sakhya Sagar Lake and 03 sites in Madhav Lake were recorded. The Landing No. 5 has 64 wetland bird species, which is highest in number, 41 at Landing No. 3, 29 species near Bhadaiya Kund, 28 species at Landing No. 4, 24 species at Landing No. 10, 22 species at near Sailing Club, 13 species at Landing No. 1, 12 species at Landing No. 2, 08 species at Landing No. 9, 3 species at Landing No. 6 and Landing No. 8 and only 2 species were observed at Landing No. 7. In the Madhav Lake the highest number of wetland species was 28 at Mound area, 19 species were counted near Pump House and 09 species were recorded at Northern Bank of the lake (Table 3.4 & Fig. 3.5).

Table 3.2: List of wetland birds with their occurrence in Sakhya Sagar and Madhav Lakes of Madhav National Park, Shivpuri, (M.P.)

S. No.	ORDER / FAMILIES Name of Species	Status	OCCURRENCES Seasonal Visitors	Sight Frequency
	PODICIPEDIFORMES / PODICIPITIDAE			
1.	Little Grebe	R	V	FC
	PELECANIFORMES / PELECANIDAE			
2.	Great White Pelican	RM	WV	U
	PHALACROCORACIDAE			
3.	Little Cormorant	RM	V	FC
4.	Great Cormorant	RM	WV	C
5.	Indian Shag	RM	WV	FC
6.	Darter or Snake Bird	RM	V	C
	CICONIIFORMES / ARDEIDAE			
7.	Indian Pond-Heron	R	ES	C
8.	Grey Heron	RM	V	FC
9.	Purple Heron	RM	WV	U
10.	Little Green Heron	R	V	FC
11.	Night Heron	R	V	U
12.	Cattle Egret	RM	ES	C
13.	Large Egret	RM	ES	C
14.	Median Egret	RM	V	FC
15.	Little Egret	R	V	FC
	CICONIIDAE			
16.	Painted Stork	RM	V	C
17.	Asian Open bill Stork	R	V	C
18.	European White Stork	M	WV	U
19.	White-necked Stork	RM	WV	FC
20.	Lesser Adjutant Stork	RM	WV	FC
21.	Black-necked Stork	R	V	FC
	THRESKIORNITHIDAE			
22.	Oriental White Ibis	R	V	U
23.	Black Ibis	R	V	U
24.	Eurasian Spoonbill	RM	WV	C
	ANSEREFORMES / ANATIDAE			
25.	Grey-leg Goose	M	WV	C
26.	Bar-headed Goose	RM	WV	C

Contd...

Table 3.2: *Contd...*

S. No.	ORDER / FAMILIES Name of Species	Status	OCCURRENCES Seasonal Visitors	Sight Frequency
27.	Brahminy Shelduck	RM	WV	C
28.	Comb-duck	RM	WV	C
29.	Lesser Whistling Duck	R	V	FC
30.	Large Whistling Duck	RM	WV	FC
31.	Northern-Pintail	M	WV	FC
32.	Gadwall	M	WV	U
33.	Shoveller	M	WV	C
34.	Common Teal	M	WV	C
35.	Red-Crested Pochard	M	WV	Rr
36.	Ferruginous Pochard	RM	WV	Rr
37.	Common Pochard	M	WV	FC
38.	Tufted Pochard	M	WV	U
39.	Mallard	RM	WV	U
40.	Eurasian Wigeon	M	WV	FC
41.	Garganey	M	WV	FC
42.	Cotton Teal	R	ES	C
	FALCONIFORMES / ACCIPITRIDAE			
43.	Marsh Harrier	M	WV	U
44.	Osprey	RM	WV	U
	GRUIFORMES / GRUIDAE			
45.	Common Crane	M	WV	Rr
46.	Sarus Crane	RM	V	Rr
	RALLIDAE			
47.	White-Breasted Waterhen	R	V	C
48.	Common Moorhen	RM	WV	C
49.	Purple Moorhen	R	V	FC
50.	Common Coot	RM	WV	C
	CHARADRIIFORMES / JACANIDAE			
51.	Pheasant-tailed Jacana	R	V	FC
52.	Bronze-Winged Jacana	R	ES	C
	CHARADRIIDAE			
53.	Red-wattled Lapwing	R	ES	C
54.	Golden Plover	RM	WV	C
55.	Little ringed Plover	M	WV	U

Contd...

Table 3.2: *Contd...*

S. No.	ORDER / FAMILIES Name of Species	Status	OCCURRENCES	
			Seasonal Visitors	Sight Frequency
56.	Spotted Redshank	M	WV	U
57.	Common Redshank	RM	WV	FC
58.	Marsh Sandpiper	M	WV	FC
59.	Common Sandpiper	RM	WV	U
60.	Jack Snipe	M	WV	U
61.	Little Stint	M	WV	FC
62.	Ruff	M	WV	U
	ROSTRATULIDAE			
63.	Painted Snipe	R	V	U
	RECURVIROSTRIDAE			
64.	Black-winged Stilt	R	ES	C
	BURHINIDAE			
65.	Stone-Curlew	R	ES	C
	GLAREOLIDAE			
66.	**Indian Courser**	**R**	**ES**	**U**
	LARIDAE			
67.	River Turn	RM	V	U
68.	Little Turn	R	V	FC
69.	Brown-headed Gull	RM	V	Rr
70.	Indian Skimmer	R	V	U
	CORACIIFORMES / ALCEDINIDAE			
71.	Lesser Pied Kingfisher	R	ES	C
72.	Small Blue Kingfisher	R	ES	C
73.	White-breasted Kingfisher	R	ES	C

Table 3.3 & Fig. 3.4: Total number and percentage of wetland birds recorded in the study area according to their occurrence

Status	Total Species	Percentage
Resident	25	34.25 %
Resident Migrant	29	39.73 %
Migrant	19	26.02 %

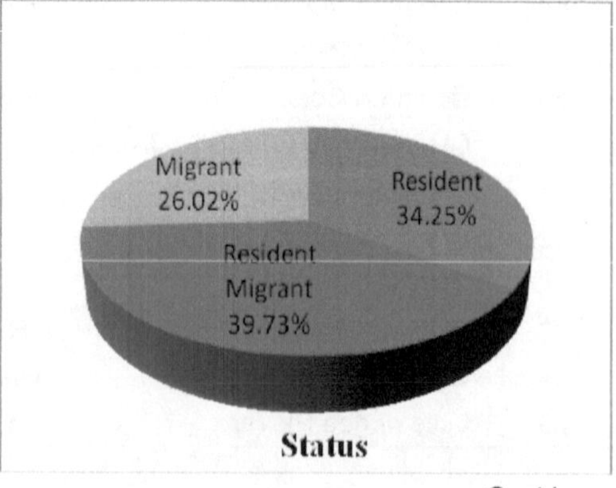

Contd...

Table 3.3 & Fig.3.4: *Contd...*

Seasonal Visitor	Total Species	Percentage
Winter Visitor	38	52.05 %
Vagrant	23	31.51 %
Every Season	12	16.44 %

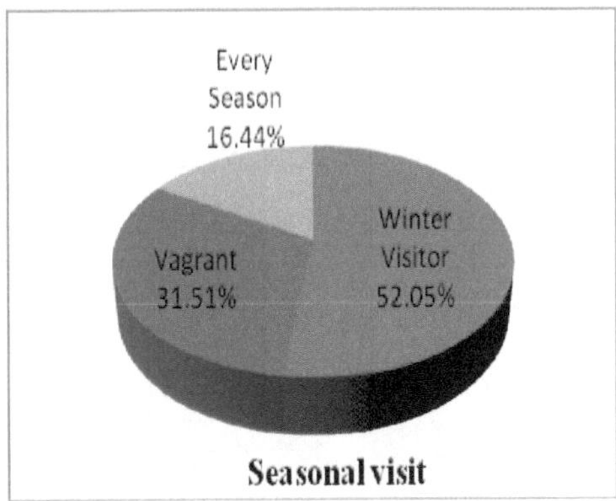

Sight Frequency	Total Species	Percentage
Common	26	35.62 %
Fairly Common	22	30.14 %
Uncommon	20	27.39 %
Rare	5	6.85 %

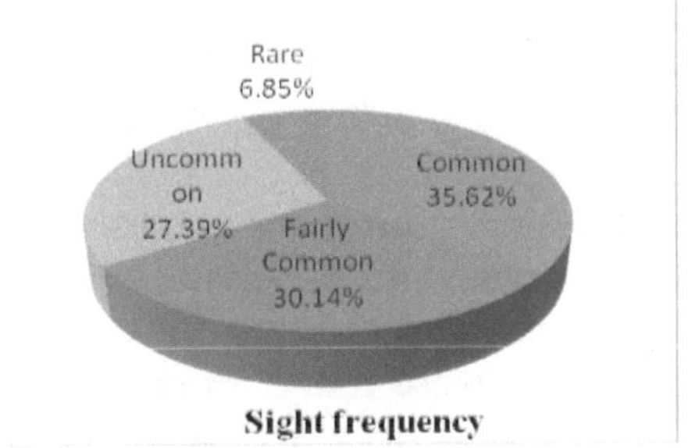

Table 3.4: Different locations utilized by migratory, resident migratory and resident birds in Sakhya Sagar Lake and Madhav Lake

S. No.	Locations		Migratory	Resident Migratory	Resident	Total
1.	Bhadaiya Kund	(BK)	2	14	13	**29**
2.	Landing No.-1	(L-1)	0	5	8	**13**
3.	Landing No.-2	(L-2)	1	8	3	**12**
4.	Landing No.-3	(L-3)	17	15	9	**41**
5.	Landing No.-4	(L-4)	9	11	8	**28**
6.	Landing No.-5	(L-5)	18	25	21	**64**
7.	Landing No.-6	(L-6)	0	1	2	**3**
8.	Landing No.-7	(L-7)	1	1	0	**2**
9.	Landing No.-8	(L-8)	0	1	2	**3**
10.	Landing No.-9	(L-9)	0	4	4	**8**
11.	Landing No.-10	(L-10)	4	10	10	**24**
12.	Sailing Club	(SC)	0	11	11	**22**
13.	Pump House	(PH)	2	6	11	**19**
14.	Mound area inside water	(MAW)	2	14	12	**28**
15.	Northern Bank of Lake	(NB)	1	3	5	**9**

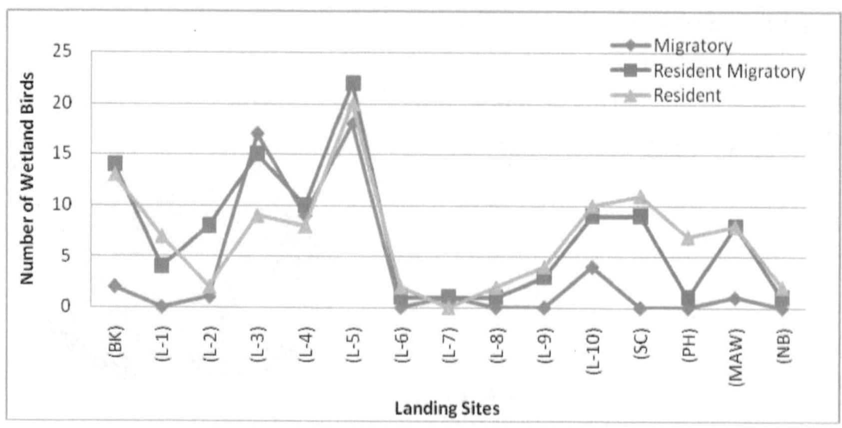

Fig. 3.5: Different locations utilized by migratory, resident migratory and resident wetland birds.

In the present study 10 categories of habitats were observed. Wetland birds were found to utilize different habitats extensively for foraging, nesting and roosting on the emergent and fringed vegetation. Some water birds require a cluster of platforms within the water bodies in order to sit for basking during the winters. On the basis of observation, 44 wetland bird species were recorded utilizing shallow marshy areas, 43 species used muddy shoreline habitat, 29 species used areas with submerged vegetation, 21 species were found in open water, 18 species in reeds and grasslands, 17 species in rocky area near water, 16 species in emergent vegetation, 11 species in floating vegetation, 08 species were found in trees and bushes standing near water bodies and 08 species were recorded flying over water bodies (Table 3.5 and Fig. 3.6).

Table 3.5: Different habitats utilized by migratory, resident migratory and resident birds in Sakhya Sagar Lake and Madhav

S. No.	Habitat		Migra-tory	Resident Migratory	Resident	Total
1.	Submerged Vegetation	(H-1)	8	11	10	**29**
2.	Emergent Vegetation	(H-2)	5	8	3	**16**
3.	Flouting Vegetation	(H-3)	0	3	8	**11**
4.	Open Water	(H-4)	8	10	3	**21**
5.	Muddy Shoreline	(H-5)	10	16	17	**43**
6.	Mound area inside water	(H-6)	15	18	11	**44**
7.	Rocky area near water	(H-7)	0	7	10	**17**
8.	Tree and Bushes standing near water	(H-8)	1	4	3	**8**
9.	Reeds and Grassland near water	(H-9)	1	7	10	**18**
10.	Fly over and around lakes	(H-10)	1	3	4	**8**

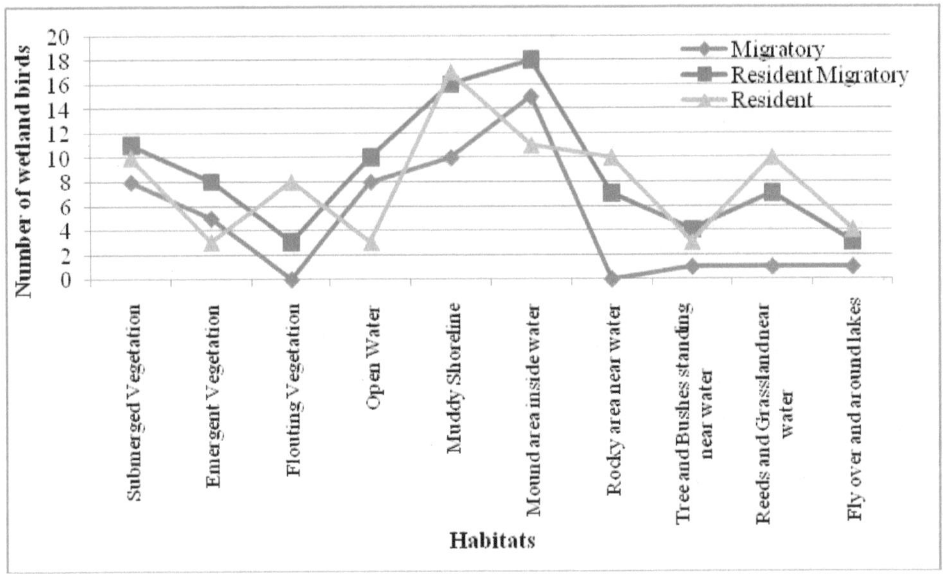

Fig. 3.6: Different habitats utilized by migratory, resident migratory and resident wetland birds.

The water bodies of Madhav National Park are very favorable habitat for number of wetland birds including residents, resident migrant and migratory species. It is because the Sakhya Sagar and Madhav Lakes are perennial water bodies which have suitable water level at the time when migratory birds approach this area. Also in the winter season there is preponderant of aquatic vegetation which provide ample food and space for them to live. In the shallow water invertebrates, fishes and amphibians are present which are favorable foods for migratory birds. In deep water also some birds prey upon fishes. In this way some birds like Pelican, Cormorants, Storks, Pochards expend their forage ground depending upon the water level of these lakes. Some species are mostly observed on the shore of the lakes like Herons, Egrets, Stilt, Jacanas, Moorhens, Waterhen, Snipe and Kingfishers. In addition to this some birds in mix groups are also observed from the banks and shallow water of the lakes like Grebe, Coot, Spoonbill, Skimmer and Little Turn. Shrivastava (1994) enlisted 230 resident and 72 migrant birds to MNP. Chandra and Nema (2006) provided a complete checklist of birds, including terrestrial and arboreal birds (239 species, 58 families) from MNP. Sharma *et al.,* (2006) studied the Mansagar Reservoir and a total of 82 bird species; 39 of waterfowl and 43 of terrestrial birds were noted. Similar kinds of observations were noted by Yahya

(1988) in the Periyar Tiger Reserve, Kerala; Kotangale and Ghosh (1997) Ranapratap Sagar, Rajasthan; Dayananda (2005) in the Gudavi Bird Sanctuary, Karnataka. Raj *et al.*, (2010) recorded total 101 species of resident and migratory in the Pallikaranai wetlands, Chennai. Shukla and Lone (2010) recorded 63 species of water birds, belonging to 17 families from Sur Sarover Lake, Agra U.P., India.

References

Ali, S. (2002). The book of Indian Birds. 13 Ed. Bombay Natural History Society. Oxford University Press.

Ali, S. & Ripley, S.D. (1987). *Compact handbook of the birds of India and Pakistan together with those of Bangladesh, Nepal, Bhutan and Sri Lanka.* Oxford University Press, Delhi.

Chandra, K. & Nema, D.K. (2006). Birds of Madhav National Park, Shivpuri, Madhya Pradesh. *Journal of Tropical Forestry, Jan. - June. Vol.* 22 (I&II): 70-78.

Dayananda, G.Y. (2005). Avifaunal Diversity of Gudavi Bird Sanctuary, Sorab, Shimog, Karnataka. *Our Nature* (2009) 7:100-109

Grimmett, R., Inskipp, C. & Inskipp, T. (1999). *Pocket guide to the birds of the Indian subcontinent.* Oxford University Press, Delhi.

Islam, M.Z. (2006). Conservation of waterbirds and wintering areas through Important Bird Areas in India. *Waterbirds around the world. Bombay Natural History Society.*The Stationery Office, Edinburgh, UK. p. 687.

Kotangale, J.P. and Ghosh, T.K. (1997). Diversity of avifauna in and around Ranapratapsagar, Rajasthan. *Encology* 12 (3), 29-37.

Raj Nikhil, P.P., Ranjini, J., Dhanya, R., Subramanian, J., Azeez, P.A. and Bhupathy, S. (2010). Consolidated checklist of birds in the Pallikaranai Wetlands, Chennai, India. *Journal of Threatened Taxa.* 2 (8): 1114-1118.

Rodgers,W.A., Panwar, H.S. & Mathur, V.B. 2000. Wildlife Protected Area Network in India: A Review, Executive Summary. Wildlife Institute of India, Dehradun.

Sharma, P.K., Sharma, Subhasini, Sharma, Shweta, Singh, P.K., Rathore, G.S., Sharma, Arti and Sharma, K. P. (2006). Avian Diversity of Mansagar Reservoir and its Environ. *Indian Journal of Environmental Sciences.* 10 (1) pp. 85-86.

Shrivastava, P. (1994). General faunal survey of Madhav National Park, Shivpuri, (M.P.) with special reference to Birds and Mammals. Ph.D.Thesis submitted to Jiwaji University, Gwalior, (M. P.).

Shukla, U. N. and Lone, Altaf Ahmad (2010). Water Birds of *Sur Sarovar* Bird Sanctuary, Agra, Uttar Pradesh. *Research Journal of Agricultural Sciences*, 1 (2): 135-139.

Sonobe, K. and Usui, S. (1993). *A field guide to the water birds of Asia.* Wild Bird Society of Japan, Tokyo.

Urfi, (2003). Breeding ecology of birds. *Resonance*: 22-32.

Yahya, H.S.A. (1988). Habitat Preference of Birds in the Periyar Tiger Reserve, Kerala. *Ind. J. Forestry.* 12: pp. 288-295.

Sustainable Development of Natural Resources and
Wildlife Conservation
Editor: **Ashwani Kumar Dubey**
Published by: **REGENCY PUBLICATIONS, NEW DELHI**

Pages **31-36**

4

Environment Cleanup through Vermi-technology Using Nirmalaya (Floral Waste)

Anil Pandey, Praveesh Bhati and Manish K. Sharma*

ABSTRACT

Waste materials generated from different resource ultimately enter into natural environment e.g., water and soil without any change. Ujjain is a holy city with various temples for pilgrims. Routinely flowers are used in the temple for the Pooja-Archana and after use dumped in the soil as such which ultimately increase the chance of contamination/pollution in and around the area. Vermi-technology is an economic, ecofriendly and employing, technique in which worms are used to produce compost. Floral waste collected, chopped, mixed with worms in a proportion and kept for some period of time for the compost preparation. Using this technology we can transform the harmful waste into useful compost. The compost is called vermicompost can be effectively applied into the field, garden, for improvement of plant health without harming the soil fertility.

Key words: *Worm, Compost, Nirmalaya.*

Introduction

Composting is a process in which microorganisms rapidly consume organic matter, using it as an energy source and converting it into carbon

P.G. Department of Microbiology, Govt. Madhav Vigyan Mahavidhyala, Ujjain.
* *Corresponding author:* E-mail: mks_ujjain@yahoo.com

dioxide, water, *microbial biomass*, heat and *compost*. Composting is the transformations of raw organic materials into biologically stable, humic substances suitable for a variety of soils and plant uses. Essentially, composting is controlled decomposition, the natural breakdown process that occurs when organic residue comes in contact with soil.

Basics of Vermicomposting

Vermicomposting is the process of using worms ("vermi" is Latin for "worm") to process organic food waste into nutrient-rich soil. Worms eat decaying food waste and produce vermicompost, a very effective soil amendment. One of the unique features of vermicompost is that during the process of conversion of various organic wastes by earthworms, many of the nutrients are changed to their available forms in order to make them easily utilizable by plants. Earthworms have been associated with human civilization as "Friends of farmers", forming an important group of soil animals that are known to improve soil productivity by enhancing the physical, chemical and biological characteristics of soil (Lee, 1985). Earthworms belonging to Phylum Annelida, Class Chaetopoda, and Order Oligochaeta occupy a unique position in animal kingdom (Kaviraj and Sharma 2003). Animal manure is a valuable resource as a soil fertilizer because it provides large amounts of macro- and micronutrients for crop growth and is a low-cost, environmentally-friendly alternative to mineral fertilizers. However, the use of manure in agriculture is being abandoned because of increasing transportation costs and environmental problems associated with the indiscriminate and inappropriately-timed application to agricultural fields (Hutchison *et al.*, 2005). Main aim of the present study was to observe the effect of flower extract for conversion by verms at P.G. Department of Microbiology, Govt. Science College, Ujjain.

Materials and Methods

This study was conducted during the period from October 2012 to January 2013. For this experiment *Eisenia foetida* (The Red Wiggler) was used and floral waste collected from different temples were used as a substrate. All flowers firstly taken from the sources and brought to the laboratory for further study.

Collection of substrate: The sample *e.g.,* cattle dung was collected from local animal husbandry (cattle houses) in and around, Ujjain (M.P.). The floral wastes material used in this experiment was collected from different temples such as Shri Mahakaleshwar and ISKON temples, due to huge amount of flowers (Nirmalya) are discharged after worship. Plastic and other waste and non degradable materials were separated through hand sorting from collected floral waste.

Preparation of material for composting: First step of the preparation of composting materials was to chopping of flowers into degradable form (Suthar and Singh, 2008). For the determination of composting quality of the worm used different combination of cow dunk and chopped flowers were mixed in the different ratio.

Analysis of compost: After preparation of vermicompost, different parameters were analyzed using various methods for the determination of compost quality.

- *Odor:* This detection was done simply by smelling (Rodale, 1960).
- **Heating:** It was detected by inserted the hand into the vermicompost at ½ feet depth.
- **Granule size:** During vermicomposting particle size of vermicompost measured with the help of scale in millimeter (mm).
- *pH variation:* pH of each sample was determined with the help of pH meter according to the method of Rebollido *et.al.,* (2008).
- *Colour: The colour of vermicompost was detected by simply looking them.*
- *Seed germination:* For this experiments three plant varieties *e.g.,* wheat, gram, and soybean were selected. Surface sterilized ten seeds of all three varieties were sown in the earthen pots in the 30:70 ratio of vermicompost: soil. Percent seed germination was calculated after seeds germination.

Results and Discussion

Deterioration of environment is a major problem facing by the world, the extensive use of chemical fertilizers/pesticides plays a key role in degradation of environmental resources, loss of soil fertility, and less agricultural yield and soil degradation are few of them (Inbar *et al.,* 1993). Vermicomposting is one of the best ways to dispose the wastes, not only due to its capacity of reducing the wastes, but also for its ability to remediate and amend the soil (Aleagha *et al.,* 2009).

Earthworms represent the major animal biomass in most terrestrial temperate ecosystems (Edwards & Bohlen, 1996). Physical parameters after preparation of vermicompost were analyzed by using various methods and results are depicted in the Table 4.1. The vermicomposting process includes two different phases regarding earthworm activity: (i) an active phase during which earthworms process the organic substrate, thereby modifying its physical state and microbial composition (Lores *et al.,* 2006), and (ii) a maturation phase marked by the displacement of the earthworms towards fresher layers of undigested substrate, during which the microorganisms take over the decomposition of the earthworm-processed substrate (Aira *et al.,* 2007; Gomez-Brandon *et al.,* 2011a). In addition, the nutrient content of the egested materials differs from that in the ingested material (Aira and

Domingaez 2008), which may enable better exploitation of resources, because of the presence of a pool of readily assimilable compounds in the earthworm casts. Therefore, the decaying organic matter in vermicomposting systems is a spatially and temporally heterogeneous matrix of organic resources with contrasting qualities that result from the different rates of degradation that occur during decomposition (Moore *et al.*, 2004).

Conclusion

Vermicompost production has many environmental impacts, some of which can be highlighted such as pollution reduction from manufacturing, collection and transportation of organic waste, pollution minimization from discharging of leachate contamination on the surface streets, pollution decreasing thanks to unpleasant odor, diminution of insects and vermin, loss of problems from waste accumulation in streets and generation centers, general environmental protection, and pollution reduction of agricultural lands used as landfill.

Figs. 4.1, 4.2, and 4.3. Floral waste and cow dunk for the preparation of vermicompost

The concept of vermiculture bio-technology gives hopes for healthy ecology and acts as a versatile natural bio-reactor. Earthworms which form one of the major soil macrofauna are of very important group of secondary decomposers. Previous reports suggested that the earthworm population size and their activities in the soil are closely related to the extent of organic matter input to the soil (Edwards and Lofty, 1977). In present study it was noticed that flower extracts prepared for the vermicompost was found to very effective when mix with a proper ratio with the soil as indicated in the Table 4.2. Apart for this it was also reported in the present study that it is best for germination of wheat and Gram (Table 4.2) compared to control where only soil was used in a pot to study seed germination. As a result, vermicompost has a potential for improving plant growth and dry matter yield when added to the soil (Atiyeh, 2000; Zaller, 2007).

Table 4.1: Physical parameters and results

S.No.	Parameters	Results
1.	Odor	*Odorness (Dung type)*
2.	pH variation	*Varied*
3.	Color	*Yellowish brown*
4.	Health of earthworms	*Average*
5.	Temperature variation	*More variated*
6.	Weight of biomass	*10% reduced from initial biomass*
7.	Moisture percentage (%)	*70-80%*

Table 4.2: Effect of vermicompost on Percent (%) Seed germination of three plant varieties (n=10)

S. No.	Experiment	Wheat seed	Gram seed	Soybean seed
1	E1 (20:80) (V:S)	70±2.5	85±3.6	73±6.2
2	E2 (40:60) (V:S)	75±4.1	78±2.7	81±4.21
3	E3 (60:40) (S:V)	63±2.4	61±3.87	59±2.8
4	C1 (with soil)	70±1.8	90±4.2	87±3.5
5	C2 (with vermicompost only)	51±2.0	48±3.3	56±4.2

References

Aira M. Monroy F. and Dominguez J. (2007). Eisenia fetida (Oligochaeta: Lumbricidae) Modifies the Structure and Physiological Capabilities of Microbial Communities Improving Carbon Mineralization During Vermicomposting of Pig Manure. *Microbial Ecology*, Vol. 54, No. 4: 662-671.

Aira M. and Dominguez J. (2008). Optimizing Vermicomposting of Animal Wastes: Effects of Dose of Manure Application on Carbon Loss and Microbial Stabilization. *Journal of Environmental Management*, vol. 88, No. 4:1525-1529.

Aira M. Monroy F. and Domínguez J. (2009). Changes in Bacterial Numbers and Microbial Activity of Pig Slurry during Gut Transit of Epigeic and Anecic Earthworms. *Journal of Hazardous Materials*, vol. 162, No. 2-3: 1404-1407.

Aleagha M.H., Pedrau H., Omani G. (2009). Bioaccumulation of heavy metals by Iranian earthworms (*Eisenia fetida*) in the process of vermicomposting. Europian *Journal Agricultural Environmental Science*. 4: 480-484.

Edwards C.A. and Bohlen P.J. (1996) Lee KE (1985). Biology and Ecology of Earthworms, 3rd Edition, Chapman and Hall London, 426 pp.

Edwards C.A and Lofty J.R. (1977). Biology of Earthworms. Chapman and Hall Ltd. London. pp. 333.

Gomez-Brandon M., Lazcano C., Lores M. and Domínguez J. (2011). Short-term stabilization of grape marc through earthworms. *Journal of Hazardous Materials*, 187: 291-295.

Gomez-Brandon M., Aira M. Lores M. and Dominguez J. (2011b). Changes in Microbial Community Structure and Function during Vermicomposting of Pig Slurry. *Bioresource Technology*, vol. 102, No. 5: 4171-4178.

Hutchison M.L., Walters L.D., Avery S.M., Munro F. and Moore A. (2005). Analyses of livestock production, waste storage, and pathogen levels and prevalence in farm manures. *Applied and Environmental Microbiology* 71:1231–1236.

Inbar Y., Hadar Y and Chen Y. (1993). Recycling of cattle manure: the composting process and characterization of maturity. *Journal of Environmental Quality*, 22: 857–863.

Kaviraj S. S. (2003). Municipal solid waste management through vermicomposting employing exotic and local species of earthworms. *Bioresource Technology*. **90:** 169-173.

Lee KE (1985). Earthworms: Their ecology and relationship with soil and land use, Academic Press, Sdyney, Australia, p. 411.

Lores M. M., Gomez-Brandon D., Perez-Diaz and Domínguez J (2006). Using FAME profiles for the characterization of animal wastes and vermicomposts. *Soil Biology and Biochemistry* 38: 2993-2996.

Melville K.S., Morin P.J., Nadelhoffer K., Rosemond A.D., Post D.M., Sabo J.L., Scow K.M., Vanni M.J. and Wall D.H. (2004). Detritus, Trophic Dynamics and Biodiversity. *Ecology Letters*, vol. 7: 584–600.

Moore J.C., Berlow E.L., Coleman D.C., de Ruiter P.C., Dong Q., Johnson N.C., McCann K.S., Melville K., Morin P.J., Nadelhoffer K., Rosemond A.D., Post D.M., Sabo J.L., Scow K.M., Vanni M.J. and Wall D.H. (2004). Detritus, Trophic Dynamics and Biodiversity. *Ecology Letters*, vol. 7, pp. 584–600.

Rebollido R., Martinez J., Aguilera Y., Melchor K., Koerner I., Stegmann R. (2008). Microbail populations during composting process of organic fraction of municipal solid waste. *Applied Ecology and Environmental Research* 6(3): 61-67.

Rodale J.I. (1960).The Complete Book of Composting, Rodale Books, Emmaus, PA.

Suthar S, Singh S (2008) Feasibility of vermicomposting in biostabilization of sludge from a distillery industry. *Science of Total Environment* 394(2–3):237–243.

Zaller J.G. (2007). Vermicompost as a substitute for peat in potting media: Effects on germination, biomass allocation, yields and fruit quality of three tomato varieties. *Scientia Horticulturae*, 112, 191-199.

Sustainable Development of Natural Resources and
Wildlife Conservation
Editor: **Ashwani Kumar Dubey**
Published by: **REGENCY PUBLICATIONS, NEW DELHI**

Pages 37-41

5

Arrival and Departure of Some Migratory Birds in Nauradehi Wildlife Sanctuary

Rekha Rai

ABSTRACT

The timing of migration is controlled primarily by changes in day length. Migrating birds navigate using celestial cues from the sun and stars, the earth's magnetic field, and probably also mental maps. Migration has developed independently in different groups of birds and does not appear to require genetic change; some birds have acquired migratory behaviour since the last ice age.

Many bird populations migrate long distances along a flyway. The most common pattern involves flying north in the spring to breed in the temperate or Arctic summer and returning in the autumn to wintering grounds in warmer regions to the south.

Birds are regulator and indicator of biological productivity. They are having Diversified habitat that indicates the abundance of life supporting systems and non-contamination of the environment by the pollutants. As top consumers, they can provide information on the major changes in the flow of the energy and health of the entire system. In the present work is related to migration of Migratory birds in winder and summer season in Nauradehi sanctuary.

Key words: *Nauradehi Sanctuary, Birds and Migratory Birds.*

Govt. Girls P.G. Excellence collage Sagar (M.P.)
Corresponding author: Email: rai.rekha90@yahoo.in

Introduction

Nauradehi Wildlife Sanctuary, is one of the largest wildlife Sanctuaries in India, is located in the districts of Sagar, Damoh and Narsinghpur of Madhya Pradesh State. Nauradehi Spreads across as area of about 1197 Sq km. The sanctuary is well known for its unique jungle safari destination in India. In addition it offers adventures of Nature trek.

The International Union of Biological Sciences in 1982 Initiated a world wide programme for indemnifying and applying biological indicators in Environmental monitoring specially evaluate the effects of hazardous substances. Thus, bio-indicator methods can be applied in predicting. well in advance to assess the impact of human activities particularly of pollutants on the living systems. The faunistic composition and community stracture of fauna of a particular inland water system may be useful in assessing the quality of the environment. three basic king of information may be necessary in the monitoring of the aquatic environment

(i) The number of taxonomic units present

(ii) The number of individuals per species and

(iii) the diversity of species present.

Materials and Methods

For the present investigation line and point transect methods were used. *Visual census method* was used for estimation of the bird population with the help of field binocular from representative points. Observations were made from 7:00 A.M. to 10:00 AM and 4:00 PM to 6:00 PM on any field day the following information was recorded (a) date (b) Place (c) Bird species (d) Number of each species. Birds were photographed with the help of cannon SLR Camera.

Results

In the present study the migratory birds of Nauradehi wildlife sanctuary were observed mostly in winter season. Some of the migratory birds were observed in Summer season while some birds were sighted only once and, therefore, their arrival and departure patterns could not be ascertained.

Mostly birds began to arrive in the Nauradehi in the month of Nov. and all of them leave by April. Arrival of *Starling* was Recorded by the month of January. The departure began from the end of the April and in some cases was completed by May The Data are showing the Period of First Appearance / abundance/ Disappearance for selected visiting species of Birds in *Nauradehi Sanctuary,* in Summer Season (Table 5.1).

Map 5.1: Showing N.P. and Sanctuary in M.P. and Location of N.W.S.

Table 5.1: List of Migratory Birds coming to Nauradehi Sanctuary in Summer Season

S.No.	Species	First appearance	Abundance	Disappearance
1	Asian Koel	Apr	Apr	Jul
2	Balck Crowned Night Heron	May	Mar	Aug
3	Blue- Cheeked Bee Eater	Apr	Apr	Aug
4	Blue tailed Bee- Eater	Mar	May	Sep
5	Comb Duck	Apr	Apr	Jul
6	Cuckoos (Indian Hawk Cuckoo or Brain Fever Bird)	May	Apr	Sep
7	Eurasian Golden Oriole	Apr	Mar	Aug

The Data are showing the Period of First Appearance / abundance/ Disappearance for selected visiting species of Birds in *Nauradehi Sanctuary*.

Table 5.2: List of Migratory Birds coming to Nauradehi Sanctuary in Winter Season

S.No.	Species	First appearance	Abundance	Disappearance
1	Black Tailed godwit	Oct.	Dec	May End
2	Black Winged Stilt	Nov.	Jan	Apr.
3	Bluethroat	Nov	Jan	Apr.
4	Common Greenshank	Nov	Nov	Apr.
5	Common Teal	Nov	Nov	Apr.
6	Eurasian Wigeon	Nov	Dec	Apr.
7	Gadwall	Nov	Dec/Jan	Apr.
8	Greater Flamingo	Nov	Mar	Apr.
9	Long Billed Pipit	Nov	Feb	May End
10	Northern Pintail	Nov. Dec	Jan	Apr.
11	Northern Shoveler	Nov	Dec	Apr.
12	Rosy Pelican	Nov	Feb	Apr.
13	Ruff	Dec	Mar	Apr.
14	Spotted Redshank	Dec	Dec	Apr.
15	Spotted Sandpiper	Nov	Feb	Apr.
16	Starling	Jan	Nov	May End
17	Siberian Cranes	Oct.	Dec	May End
18	White wagtail	Nov.	Jan	Apr.
19	Wood Sandpiper	Nov	Jan	Apr.
20	Yellow wagtail	Nov	Nov	Apr.

The Data are showing the Period of First Appearance / abundance/ Disappearance for selected visiting species of Birds in *Nauradehi Sanctuary.*

Some birds began to arrive in the Nauradehi in the month of April. and all of them leave by August. Arrival of *Cuckoos (Indian Hawk Cuckoo or Brain Fever Bird* was Recorded by the month of *September*. The Data are showing the Period of First Appearance / abundance/ Disappearance for selected visiting species of Birds in *Nauradehi Sanctuary,* in Summer Season (Table 5.2).

Pandit (1991) has recorded the King vulture (*Sarcogyps calvus*) in Andhra Pradesh in January and shree kumar and gupta (2009) noted it in Kerala in December.

According to Thomas (1982) the Swallow (Hirundo rustica) visits kerala in a large number whose as this species is widespread in NCS from october to March. this type of grouping of birds depending upon relative use of the wetland habitat is also supported by Ali and Vijayan (1985) and Kaufmann and Cawley (1986).

In many migratory birds, mates retain throughout the season for successful breeding (Shah 1984) Wicox (1986) and Hunt (1980) have also reported that in sea birds including terns. Monogamy was predomainant and both sexes play active role throughout the breeding season. It has also been reported that widespread occurrence of monogamy among birds occurs because reproductive oppotutnities may be severaly restricted by the availability of potential mates (Gochfild 1980).

References

1. Ali, S. and L. Futehally (1967). Common birds. National Book Trust, New Delhi, India.
2. Ali, S. and V.S. Vijyan (1985). Ecolony and Keoladeo National Park, Summary report of 1980-1985. Bombay Nat. Hist. Soc, Research Report, Hydro. 2.
3. Ali, S. and V.S. Vijyan (1985). Ecology of Keoladeo National Park, Summary Report of 1980-1985. Bombay Nat. Hist. Soc, Research Report, Hydro. 2
4. Gochfeld, M. (1970). Mechanism, and adaptive value of reporductive synchrony in colonical sea birds. Behaviour of Marine animals and marine birds Eds. banger, *J. Olla, B.L.* and *winn, H.E.* 207-270 Plenum Press, New York.
5. Hunt, G.L. J2 (1980). Mate selection and mating system in sea birds. In Banger, J.BL Olla, and L.H.E. Winn (eds). Behavior of Marine animals 4, Marine Birds, New York: Plenum Press.
6. Kaufman, G.W. and E. Cawley, (1986). Nesting sucess and feeding habits of great blue herons on the Mississippi river near cassville, wisconsin and east dubugue, Illinois, *Proc. 10 WA Acad. Sci*, 93 (4) 84-87.
7. Kumar P. and S.K. Gupta (2009). Diversity and abundance of wetland birds around Kurukshetra, India. *Our nature* 7:212-217.
8. Pandit A.K. (1991). Conservation of wildlife resources in wetland ecosystems of kashmir, India, *Journal of Environmental Management* 33 (2) 143-154.
9. Thomas G.J. (1982) Autumn and winter feeding ecolony of waterfowl at the ouse washes, England. *J. Zool. Lond* 197: 131-172.
10. Wicox C.G. (1986). Comparison of shore bird and waterfowl densities on resorted and natural intertidal mudflats at upper new port Bay, California, USA. Colonial Water Bird 9: 218-226.

Sustainable Development of Natural Resources and   *Pages* **42-47**
Wildlife Conservation
Editor: **Ashwani Kumar Dubey**
Published by: **REGENCY PUBLICATIONS, NEW DELHI**

6

Toxicity Effect of Plant Extract on Egg Masses of *Lymnaea stagnalis*

Payal Mahobiya and Mangla Bhide

ABSTRACT

Lymnaea stagnalis is an herbivorous snail and common pest of aquatic plants and also a prolific breeder. It is commonly known as pond snail and found in all types of fresh water ponds and lake. In the present investigation glycoside of Cerbera thevetia was tested for lethal toxicity against Gyraulus convexiusculus. In the present investigation has also been taken In the present study, spawn of Lymnaea stagnalis was exposed to different concentration of plant extracted of Cerbera thevetia, they showed hyper-irritability manifested in climbing behavior at the surface of the water avoid contact with treated medium and to take in fresh water. The egg masses were swollen and were sticky then in controls. Snails are pest and prolific. So it is very essential to control the population from glycocides of plant extract of Cerbera thevetia.

Introduction

Lymnaea stagnalis snail is the most harmful pest of paddy crops and aquatic vegetation. Most of workers have investigated the toxicity of pesticides for land snails. Most of the study about snails development were around or after hatching. In the present investigation has also been taken

Assistant Professor, Department of Zoology, Dr. H.S. Gour (Central) University, Sagar. (M.P.)
Corresponding author: Email: 1607payal@gmail.com

to study the effect of glycosides extracted of *Cerbera thevetia* on the egg masses of *Lymnaea stagnalis*. The embryos of *Lymnaea stagnalis* are well visible inside the egg capsules and therefore this snail is very easy ideal material to study the embryonic development.

The work on the glycoside toxicity in *Lymnaea stagnalis* is yet scanty, so this is attempt was performed to evaluate of glycoside on the toxicity on the different developmental stages of *Lymnaea stagnalis*.

Materials and Methods

Adult snails of Lymnaea stagnalis were collected from their natural habitats in Sagar pond. The snails were kept in glass container filled of water. The snails were fed on fresh aquatic vegetation. The collected snails were acclimatized for 7 days under laboratory conditions according to method adopted by Subbarao (1989). The water was replaced with fresh water 3 times in a week. They The number of eggs sacs and the eggs per sacs were counted for each snail. The egg masses laid by these snails were introduced via to different concentration concentrations of plants seed extracts and the data was collected in triplicate and calculated the values of LC_{100}, LC_{50}, LC_0 and sub-lethal concentration were detected out for each group separately and data was summarized in table no.1(Probit analysis method adopted after Finney, 1971). Each egg masses contain about 30 egg capsules. Different developmental stages of *Lymnaea stagnalis* selected in the present investigation as follows:

1. Cleavage
2. Blastula
3. Gastrula
4. Post gastrular changes e.g. Morphogenesis and Organogenesis.
5. Formation of trochophore larvae.
6. Number of trochophore transformed into veliger larvae.
7. Torsion in veliger larvae.
8. Metamorphosis of veliger into young snails.
9. Hatching of young snails from their respective egg capsules.

The seeds of *Cerbera thevetia* have been analysed to find out the glycosides.

The defatted seed powder (100 gm) of each plant was extracted with 95% ethanol for 20 hrs separately in a Soxhlet apparatus. The ethanolic extract on concentration gave a brownish syrupy mass. It was then successively extracted with acetone. The process of dissolving and precipitation was repeated several times and finally through a bed of activated charcoal. On removal of the solvent a brownish mass of glycoside was obtained in a yield of 6.83%.

Results

Cleavage is spiral, begins about 2½ to 3 hours after the eggs masses were laid in the control group of *Lymnaea stagnalis* but it started after 6±2 hours after treatment with glycoside in *Lymnaea stagnalis*. The dose and duration of treatment dependent increase in the duration of cleavage has been observed in the experimental snails of *Lymnaea stagnalis*.

In the present investigation in *Lymnaea stagnalis* it was observed that blastula period was increased in paclitaxel treatment. However, mortality during blastula stage ranges from 0.5 to 1.0 per cent. The increase in the duration of blastula was more or less same in the experimental snails of *Lymnaea stagnalis*.

In the present investigation in *Lymnaea stagnalis* the gastrulation period increased by 5±2 hours in treatment. This stage was found to be more susceptible as mortality occurred in the later gastrula stages. After gastrulation three germinal layers were formed and organogenesis and morphogenesis were started in control as well as in experimental snails but the duration was somewhat more prolonged in experimental snails and was found to be dependent on dose and duration of treatment in comparison to the control groups.

The young snail hatchability percentage was very low with high percentage of mortality after high concentration of glycoside treatment in comparison to control. Data on the morphological observation on different survivality percentage, mortality percentage and hatchability of young snails was recorded and summarized in table 6.2.

Discussion

In the present investigation egg laying was observed throughout the year in *Lymnaea stagnalis* and it was observed that egg masses were laid in large number from July to October as also reported by Gupta (2003), Nema (2005) Jain (2007) and Mahobiya *et al* (2012).

In the present investigation after treatment with glycocide as also observed by Grasveld (1949) in *Lymnaea stagnalis*. Isolated blastomeres generally give rise to the same sequence of cells as in normal development, while in the present investigation blastomeres were not isolated. In further development the micromeres spread over the macromeres. This is followed by slight invagination of the macromeres resulting in a small depression representing a rudimentary archentric cavity and blastopore as also observed in the present investigation in *Lymnaea stagnalis*. The endodermis cells originated from the macromeres and fill the whole of the interior of the embryos. The gastrula is therefore a sterogastrula.

Two larval stages *e.g.* trochophore and veliger larval stages were found in the development of *Lymnaea stagnalis* as also observed by Gupta (2003) in *Lymnaea* spp. and *Gyraulus* spp., Nema (2005) in *Lymnaea* spp., Jain (2007) in pulmonates and Mahobiya *et al* (2012) in Lymnaea spp. while in the present investigation larval development arrest has been observed in large number of egg capsules in *Lymnaea stagnalis* due to the intoxication of alkaloid and the effect was more pronounced in paclitaxel treatment.

In the present investigation in *Lymnaea stagnalis* was observed that the snails developed from the treated egg masses had thin and transparent shell due to decalcification as also observed by Bhide (1991) in *Lymnaea stagnalis* after nuvan and methyl parathion exposure, Gupta (2003) in *Lymnaea stagnalis* and *Gyraulus convexiusculus* after some pesticides and dye treatment, Nema (2005) in *Lymnaea* spp. and Jain (2007) in *Lymnaea* spp. and *Gyraulus* spp. after treatment with some pesticides respectively.

It could be concluded from the present investigation that the glycocide used for the treatment, were able to arrest the development at any stage, in which larval stages were most susceptible suggested the larvicidal nature of the glycoside which induces teratogenicity.

Table 6.1: Data on Toxicity of Plant Extract of *Cerbera thevetia* on Egg Masses of *Lymnaea stagnalis*.

S.No.	Name of the Glycoside compound	Concentration of the Glycocide Compound	Duration (hrs.)	Mortality (%)	Lethal conc. value
1.	Plant Extract of *Cerbera thevetia*	1.0%	72	100%	LC_{100}
2.		0.5%	72	50%	LC_{50}
3.		0.26%	72	Nil	LC_0
4.		0.25%	72	Nil	Sublethal concentration

Results

0.25% concentration of colchicine was considered as sublethal concentration value.

Table 6.2: Developmental Data of *Lymnaea stagnalis* under the influence of cerberin (Methanolic Extract)

Kind of the Glycoside	Conc. of the Glycoside	Total No. of Egg Capsules	No. of eggs completed cleavage	No. of eggs completed blastula	No. of eggs completed gastrula	No. of trochophore formed	No. of veliger formed	No. of veliger completed torsion	Total No. of young snails hatched	No. of young snails survived upto adulthood
Control	No. of trace of any Glycoside	50	49 ± 1	49 ± 1	49 ± 1	48 ± 1	48 ± 1	47 ± 1	48 ± 2	47 ± 1
	1.0%	50	48 ± 1	47 ± 2	38 ± 1	32 ± 1	25 ± 2	24 ± 1	22 ± 2	19 ± 2
Plant Extract of Cerbera thevetia	0.5%	50	48 ± 1	47 ± 2	25 ± 2	24 ± 1	22 ± 2	19 ± 2	15 ± 2	15 ± 1
	0.26%	50	48 ± 1	47 ± 2	24 ± 1	19 ± 2	18 ± 2	16 ± 2	14 ± 1	11 ± 2
	0.25%	50	48 ± 1	47 ± 2	20 ± 2	15 ± 2	13 ± 2	11 ± 2	11 ± 1	9 ± 1

References

1. Bhide, M. (1991). Effect of organophosphorus pesticides on the behavior, mortality and on the development of *Lymnaea stagnalis*. Perspectives in Aquatic Eco. Toxicology Ltd., (Ed. Nalin K Shastree), Narandra Publishing House, Delhi, India.

2. Finney, D.J. (1971). Probit analysis. 3rd Edn. Cambridge University Press, London.

3. Grasveld, M. S. (1949). On the influence of various chlorides on maturation and cleavage of the egg of *Lymnaea stagnalis* L. *Proc. Koninkl Ned. Akad.*, Wetens Chap., 52, 284.

4. Gupta, P. (2003). Effect of some toxicants on the development of some freshwater snails Ph.D. Thesis, Dr. H. S. Gour Vishwavidyalaya, Sagar (M.P.).

5. Jain, S. (2007). Effect of molluscicides of plant origin on the reproductive performance of *Lymnaea* spp. and *Gyralulus* spp. Ph. D. Thesis, Dr. H. S. Gour University, Sagar (M.P.).

6. Mahobiya, P., Shilpi, K., Tomar, D. and Bhide, M. (2012). Effect of colchicine on different developmental stages of *Lymnaea stagnalis*, *Journal of Environment and Bio-Sciences*, Vol 26, 57-58.

7. Nema, P. (2005). Effect of some plant extracts on the development of *Lymnaea* spp. Ph.D. Thesis, Dr. H.S. Gour University, Sagar (M.P.).

8. Subbarao, N.V. (1989). Handbook of fresh water molluscs of India. Radiant Process Pvt. Ltd. Calcutta, India.

Sustainable Development of Natural Resources and   *Pages* **48-53**
Wildlife Conservation
Editor: **Ashwani Kumar Dubey**
Published by: **REGENCY PUBLICATIONS, NEW DELHI**

7

Impact of Waste Water Discharge on Water Quality of River Mandakin at Chitrakoot Satna, M.P., India

Sadhana Chaurasia* and Raj Karan**

ABSTRACT

The detailed survey of river revealed that small areas as well as large areas which fall in way of river, dump domestic and toxic wastes in the river. Domestic wastewater, agricultural runoffs, mass bathing, offering of religious materials, clay idols, etc. increases the pollution in water. River water samples were analyzed in terms of physical and chemical water quality parameters. Physico-chemical quality of river was found very poor. The quality of water indicated that river seemed much polluted from Arogyadham to Karwi Bridge due to several drains joining the river as point pollution sources. The point sources discharge domestic sewage or wastewater through open drains, sewerage systems etc. BOD and COD were found in higher range. The river water was highly polluted with reference to physico-chemical quality and biological quality. Increased pollution load deteriorating the water quality of river Mandakini day by day. The results of the study showed that despite of all awareness efforts pollution load is still increasing making water unfit for drinking and human consumption without proper treatment.

---

*Head, Dept. of Energy & Environment, MGCGV, Chitrakoot, Satna-485331, M.P.
**Research Scholar, Dept. of Energy & Environment, MGCGV, Chitrakoot, Satna-485331 M.P.
Corresponding author: Email: sadhanamgcgv@gmail.com, surya.raj80@gmail.com

Introduction

The river Mandakini originates from the hills of Khillora near Pindra village, Majhagawan block (25° 09'24.8" N, 80° 52' 55.3"E), Satna district, M.P. at an elevation of 156 m above mean Sea level in the state of Madhya Pradesh of northern India. Whole watershed area is 1956.3 sq. km. The river shared between M.P. and U.P. portions. Sati Anusuiya is a perennial reach of Mandakini River where a large number of small springs feed the river. The number of drains carrying wastewater of Chitrakoot localities joining the river at various points increasing the pollution load of the river and altering its water quality (Chaurasia & Raj Karan, 2013). Water sources available for drinking and other domestic purpose must possess high degree of purity, free from chemical contamination and micro-organism (Borul, *et al.*, 2012).

In India there is enormous number of natural and manmade water bodies used for various purposes, mainly for drinking and agriculture. However, in recent years due to rapid urbanization and modern agricultural activities, the quality of water bodies deteriorated causing environmental hazards (Bhadja & Vaghela, 2013).

Materials and Methods

On the stretch of river (4.20 km approximate) three sampling station were selected. All sampling stations namely Arogyadham Jankikund and Ramghat were in M.P. part. The water quality was studied for one year for various physico-chemical parameters. All the analysis was done as per APHA-AWWA-WPCF 2005.

The selected sampling stations could be categorized as given below:
(a) On the sources of water supply for the bathing and drinking purpose.
(b) On the polluted and relatively unpolluted stretch of the river.
(c) On the major waste water discharge from domestics and localities joining drains in to the river.

Results and Discussion

The results of the physico-chemical and biological parameters of river water are given table 7.1 and drains water are given table 7.2.

Table 7.1: Physico-chemical characteristics of river at selected stations (2013).

Sampling Station	pH	Turbidity (NTU)	TDS (mg/l)	E.C. (µmho /cm)	DO (mg/l)	BOD (mg/l)	COD (mg/l)	F.Coliform (MPN/100 ml)
Arogyadham	8.12	14.32	289.23	362.56	7.88	7.32	10.43	842
Jankikund	8.13	18.28	360.27	373.67	7.72	12.12	19.76	920
Ramghat	8.16	28.11	392.28	482.82	5.98	28.04	44.87	2432

Table 7.2: Physico-chemical characteristics of drains joining the river (2013).

Sampling Station	pH	Turbidity (NTU)	TDS (mg/l)	E.C. (µmho /cm)	DO	BOD (mg/l)	COD (mg/l)	F.Coliform (MPN/100 ml)
D1. Arogyadham	7.74	39.23	587.78	964.43	3.85	60.89	123.07	3654
D2 Jankikund	7.76	42.78	627.09	942.23	2.78	63.45	132.84	4172
D3 Ramghat	8.00	48.92	743.98	1112.65	1.23	69.04	140.17	4834

*D1, D2 & D3= Drain (Near Sampling Station)

pH

pH range of 6.0 to 9.0 appears to provide protection for the life of fresh water fish and bottom dwelling invertebrates. The pH value of river was recorded between 8.12-8.16. Minimum pH of river water was observed 8.12 of Arogyadham while maximum 8.16 at Ramghat.

The pH value of drains was recorded in the range 7.74-8.00. The minimum value was 7.74 at D1 Arogyadham while maximum value was 8.00 at D3 Ramghat.

Turbidity

Turbidity in wastewater is contribute by various organic and inorganic materials such as fecal matter, pieces of paper cigarette-ends match sticks, greases, fruit skins, etc. The turbidity value of river was found in the range 14.32-28.11 NTU. The minimum value was found 14.32 NTU at Arogyadham while maximum value was 28.11 NTU at Ramghat.

The turbidity value of drain was found in the range 39.23-48.92. The minimum value was found 39.23 at D1 Arogyadham while maximum value was 48.92 NTU at D3 Ramghat.

Total Dissolved Solid (TDS)

Normally wastewater water contains 99.9% of water and 0.1% solid in the form of suspended and dissolved solids. Those solid which are dissolved in the sewage are called dissolved solid. TDS value of river was found in the range 289.23-392.28 mg/l. The minimum value was found 289.23 mg/l at Arogyadham while maximum value was 392.28 mg/l at Ramghat.

TDS value of drains was found in the range 587.78-743.98 mg/l. The minimum value was found 587.78 mg/l at D1Arogyadham while maximum value was 743.98 mg/l at D3 Ramghat.

Electrical Conductivity (E.C.)

The EC value of river was found in the range 362.56 to 482.82 µmho/cm. Minimum value of EC was observed 362.56 µmho/cm at Arogyadham while maximum 482.82 at Ramghat. Conductivity is widely used to indicate the total ionized constituents of water (Chaurasia and Raj Karan, 2013).

The EC value was found in the range 964.43-1112.65 µmho/cm. The minimum value was found 964.43µmho/cm at D1 Arogyadham while maximum value was 1112.65µmho/cm at D3 Ramghat.

Dissolved Oxygen (DO)

In river stretch the value of dissolved oxygen (DO) was ranged from 5.98-7.88 mg/l. Minimum value of DO was found 5.98 mg/l at Ramghat and maximum value 7.88 mg/l at Arogyadham.

The DO value of drain was found in the range 1.23-3.85 mg/l. The minimum value was found 1.23 mg/l at D3 Ramghat while maximum value was 3.85 mg/l at D1 Arogyadham. Low level of DO is again indicative of polluted nature of water body.

Biochemical Oxygen Demand (BOD)

In river stretch BOD was ranged from 7.32-28.04 mg/l. Minimum value was 7.32 mg/l at Arogyadham while maximum value was 28.04 mg/l at Ramghat.

The BOD value of drain was found in the range 60.89-69.04 mg/l. Minimum value was found 60.89 mg/l. at D1 Arogyadham while maximum value was 69.04 mg/l. Increase in BOD is due to domestic sewage, animal and crop wastes, heavy sediment, organic matter and domestic sewage that directly discharged in to river Mandakini. Similar value found by Chaurasia, 1994.

Chemical Oxygen Demand (COD)

The value of COD for river water was found in the range of 10.43-44.87 mg/l. Minimum value was 10.43 mg/l at Arogyadham while maximum value was 44.87 mg/l at Ramghat.

The COD value of drain was found in the range 123.07-140.17 mg/l. The minimum value was found 123.07 mg/l at D1 Arogyadham while maximum value was 140.17 mg/l at D3 Ramghat. Chemical Oxygen Demand is a measure of the oxidation of reduced chemicals in water. It is commonly used to indirectly measure the amount of organic compounds in water (Kumar *et al.*, 2011).

Fecal Coliform (F.C.)

The presence of fecal coliform bacteria in aquatic environmental indicates that the water has been contaminated with the fecal material of man or other animals. The value of fecal coliform of the river was found in range of 842-2432 MPN/100 ml. The minimum value was found 842 MPN/ 100 ml. at Arogyadham while maximum was found 2432 MPN/100 ml. at Ramghat.

The value of Fecal coliform of drains was found in the range of 3654-4834 MPN/100 ml. The minimum value was found 3654 MPN/100 ml at D1 Arogyadham while maximum value was 4834 MPN/100 ml at D3 Ramghat. The presence of fecal contamination is an indicator that a potential health risk exists for individuals exposed to this water.

Table 7.3: WHO standard for drinking water 1993

S.No.	Parameters	Standard
1.	pH	7.0-8.0
2.	Turbidity (NTU)	5.0
3.	Total Dissolved Solid (mg/l)	500
4.	Electrical Conductivity (μmho/cm)	300
5.	Dissolved Oxygen (mg/l)	5.0
6.	Biochemical Oxygen Demand (mg/l)	6.0
7.	Chemical Oxygen Demand (mg/l)	10
8.	Fecal coliform MPN/100ml	1000 MPN/100ml

Table 7.4: Standard limit to discharge waste water

S.No.	Parameters	Permissible limit to be discharge for irrigation	Permissible limit to be discharge to public sewers	Permissible limit to be discharge in to water sources
1.	pH	5.5-9.0	5.5-9.0	5.5-9.0
2.	BOD	100	350	30
3.	COD	250	600	250
4.	TDS	200	600	100

Source: APHA AWWA WPCF 1992.

Conclusion

From the results it was concluded that physico-chemical quality (pH, TDS, Turbidity, EC, BOD, COD, F.Coliform) of river water found beyond the permissible limit at Jankikund & Ramghat prescribed by WHO. Physico-chemical qualities of drains water were very poor as these drains carry waste in to river. The quality of drains waste water showed that the drains water discharge after purification in to the river. The river water is highly polluted due to joining of drains at river. There is a need to develop awareness among people regarding the consequence of river pollution. Wastewater discharges from several points and non-point sources (drains, domestic, agriculture etc.) in the Mandakini River caused water quality decline along the length of the Mandakini River. However, despite of self-purification capacity of the Mandakini River resulted deteriorating in the quality of River water.

Unfortunately, this study showed that the water quality of the Mandakini River is polluted, probably due to wastewater discharges as point and no-point sources. An integrated programme to manage water quality along the Mandakini River is required if the resource is not to be further degraded.

References

1. APHA-AWWA-WPCF (2005). Standard methods for the Examination of Water and Wastewater, Editor A.D. Eaton, 18th ed., American Public Health Association, Washington.

2. Bhadja, P., and Vaghela, A. (2013). Status of River water quality of Saurashtra, Gujarat, India. *International Journal of Advanced Biological Research.* 3(2): 276-280.

3. Borul, S. B., and Banmeru, P.K. (2012). Physio-chemical analysis of ground water for drinking fromselected sample points around the Manmeru Science College, Lonar Buldana district of Maharashtra, *Journal of Chemical and Pharmaceutical Research,* 4(5): 2603-2606.

4. Chaurasia, S. (1994). Water Pollution from Mass Bath in River Mandakini during Chitrakoot Deepawali Mela-1993, *Indian J. Environ. Prot.,* 14, 758-765.

5. Chaurasia, S. and Raj Karan (2013). Pollution sources and water quality of river Mandakini at Chitrakoot, *IJEP,* 33(12) 669-977.

6. Chaurasia, S. and Raj Karan (2013). Water Quality and Pollution load of River Mandakini at Chitrakoot, India" *Int. Res. J. Environment Sci.,* Vol. 2(6), 1-5.

7. Kumar V., Arya V., Dhaka A., Minakshi & Chanchal (2011). A study on physico-chemical characteristics of Yamuna River around Hamirpur (UP), Bundelkhand region central India, *International Multidisciplinary Research Journal,* 1 pp. 14-16.

Sustainable Development of Natural Resources and
Wildlife Conservation

Pages **54-60**

Editor: **Ashwani Kumar Dubey**
Published by: **REGENCY PUBLICATIONS, NEW DELHI**

8

Effect of Different Spawns and Substrates on Growth and Yield of *Pleurotus sajor-caju*

Poonam Dehariya* and Deepak Vyas

ABSTRACT

A study was conducted to examine the effect of different types of spawns on oyster mushroom (*Pleurotus sajor- caju*) production using three types of substrates conventional [Soybean straw (SS), Wheat straw (WS) and Paddy straw (PS)] and non conventional [Domestic wastes (DW), Fruit waste (FW) and Used Tea leaves (UTL)]. Locally available grains of wheat (*Triticum aestivum*), sorghum (*Sorghum vulgare*), jowar were used for spawn production. These spawns were used for spawning on the conventional and non-conventional substrates. Various parameters are examined such as spawn run time (mycelia development), pinhead formation, fruit body formation and yield. The experiments were setup as a randomized design with three replicates. Results revealed that wheat grain spawn produced better results in comparison to spawn grown on the maize and sorghum for spawn running, pinhead formation, fruit body formation and increased yield. The quickest spawn running of 17 days, early pinhead formation 21 days, greater yield of 934.4g/kg with 93.4% BE was recorded with WS as a conventional substrate. Among the non conventional substrate DW was found to be best.

Key words: *Spawn, Substrate, Grain, Conventional, Non Conventional*

Lab of Microbial Technology and Plant Pathology, Department of Botany, Dr HS Gour University, Sagar (MP).
Corresponding author: E-mail: poonam.dehariya@yahoo.com

Introduction

Thus spawn comprises mycelium of the mushroom and a supporting medium which provides nutrition to the fungus during its growth. The propagating material used by the mushroom growers for planting beds is called spawn. The spawn is equivalent to vegetative seed of higher plants (Pathak *et al.*, 2000). In mushroom growing technology, the inoculums is known as the 'spawn'. Spawn is a medium that is impregnated with mycelium made from a pure culture of the chosen mushroom strain. Spawn production is a fermentation process in which the mushroom mycelium will be increased by growing through a solid organic matrix under controlled environmental condition. In almost all cases the organic matrix will be sterilized grain *e.g.* wheat, maize, sorghum etc (Jain and Vyas, 2005; Jain, 2005).

Growing medium of the mushroom is generally known as substrate. The substrates used for cultivation of oyster mushroom are normally nitrogen deficient. An addition of organic and inorganic supplements to the substrate from outside to improve the yield of mushroom have therefore been recommended by many workers (Royse and Schisler, 1987a, 1987b; Royse and Bahler, 1988; Royse, 2002 and Madhusudhanan and Chandra Mohan, 2002; Jain and Vyas, 2002; Jain and Vyas, 2005; Chaubey *et al.*, 2010). An attractive feature of oyster mushrooms is that they can utilize a large variety of agricultural waste products and transform the lignocelluloses biomass in to high quality food, flavor and nutritive value (Quimio, 1978; Bano and Rajarathanam, 1982; Dehariya and Vyas, 2013). Cultivation of oyster mushroom on Soybean straw and other conventional substrates is cheaper result the reduction in production cost of mushroom and utilizing agric waste would certain help to reduce the environmental problems particularly accumulation of filth carbon sequesters, nutrients, metal sequestration and ultimately mushroom cultivation help us to achieve bioremediation (Dehariya *et al,* 2010).

Materials and Methods

The spawn production was carried out at the Deptt. of Botany, Dr. H.S.G. University, Sagar (M.P.) Experiments were carried out in a Lab of Microbial Technology and Plant Pathology. Stock pure culture of *P. sajor caju* obtained from JNKVV, Jabalpur (M.P.) was maintained on Potato Dextrose Agar (PDA).

Spawn Preparation

The grains, wheat, maize and sorghum were cleaned manually. The cleaned grains were thoroughly washed and soaked. Thereafter, the soaked grains were drained and the excess water removed, then additives like

Chalk powder ($CaCO_3$) at the rate of 2%, and Zypsum ($CaSO_4$) at the rate of 0.2% on dry weight basis of the grains were added (Jain, 2005). Calcium carbonate adjust the pH and Zypsum prevent the sticking of grains. The grain substrate was filled in to polypropylene bags. About 200 g of substrate was packed in each bag. The bags were sealed using cotton wool plugged conduit/poly vinyl chloride pipe rings, and covered by a piece of paper by tying a rubber band around the neck. The bags were autoclaved at 121°C, 15 psi, for 30 min and the sterilized bags were allowed to cool for 24 hours. The bags were immediately inoculated with mycelial culture of *P. sajor- caju* maintained on PDA.

Substrate preparation

A medium was prepared using conventional *viz.* soybean straw (SS), wheat straw (WS), paddy straw (PS) and non-conventional substrates *viz.* Domestic wastes (DW), Fruit waste (FW) and used tea leaves (UTL). All the test substrates were washed in fresh water. The chopped straw substrates were steeped in water containing 75 ppm carbendazim + 500ppm formaldehyde for 18 hours (Jain, 2005) for preventing mould infestation due to various other competing fungi. Spawning was done @ 2% wet weight basis of substrate by thoroughly mixing. Spawned substrate is filled up in perforated polythene bags (60 × 40 cm). Three replicates were maintained for each substrates. These bags were transferred to crop room for spawn run. For spawn run temperature and relative humidity were maintained between 30-32°C and 80-90% respectively. Average values for spawn run, pin head appearance and total yield was recorded.

Experimental Design

In the experiments complete randomized design with three replicates of each., wheat spawn, maize spawn and sorghum spawn, in six types of substrates.

Data Analysis

Data were analyzed using the analysis of variance (ANOVA) procedure by SYSTAT 12.

Results

The results of the present study are summarized and inlet in the tables (1,2,3). Table 8.1 shows growth and yield performance of *Pleurotus sajor-caju* using spawn grown on wheat grain on conventional sources (SS, WS, PS). Among the conventional substrates used, SS gave significantly higher yield (934.4gm/kg). Similarly among the non conventional substrates (DW, FW, UTL) DW was found best (718.4gm/kg). Table 8.2 shows results in the same substrates using spawn grown on sorghum grain. Here, again SS

(896.7gm/kg) was best among the conventional and DW (690.0gm/kg) was best among the non conventional substrates. Data summarized in table-3 suggest that spawn grown on jowar grain produces better result with SS (870.0gm/kg) and FW (680.0gm/kg). Irrespective of substrates used for production of *Pleurotus sajor-caju* under the cultivation condition test mushroom utilizes wheat grain more efficiently in comparison to maize grain and sorghum grain.

Table 8.1: Effect of wheat grain spawn on the growth and yield of *P. sajor-caju* on conventional and non-conventional substrates.

Substrates	Spawn run (day)	Pin head (day)	Stipe length (cm)	Cap diameter (cm)	Total yield (gm/ kg)	BE (%)
Conventional						
SS	17.0	21.0	2.7	7.8	934.4	93.4
WS	19.0	23.4	3.0	8.6	800.0	80.0
PS	21.0	24.7	2.8	8.1	750.0	75.0
SEm (±)	0.63	0.66	0.22	0.07	15.24	1.5
CD (0.05%)	2.20	2.30	0.76	0.26	52.7	5.2
Non-conventional						
DW	24.7	28.4	2.5	6.4	718.4	71.8
FW	30.4	35.0	2.7	5.1	635.0	63.5
UTL	25.7	31.4	2.6	5.7	655.0	65.5
SEm (±)	0.57	0.79	0.05	0.11	7.51	0.7
CD (0.05%)	1.99	2.74	0.17	0.39	27.32	2.7

Values are given in average of three replicates

Table 8.2: Effect of sorghum grain spawn on the growth and yield of *P. sajor-caju* on conventional and non-conventional substrates.

Substrates	Spawn run (day)	Pin head (day)	Stipe length (cm)	Cap diameter (cm)	Total yield (gm/ kg)	BE (%)
Conventional						
SS	18.7	22.0	2.7	7.6	896.7	89.6
WS	20.0	24.33	2.8	8.5	780.0	78.0
PS	22.0	27.0	2.7	7.9	720.0	72.0
SEm (±)	0.50	0.69	0.06	0.06	6.93	0.6
CD (0.05%)	1.76	2.40	0.21	2.36	24.0	2.40
Non-conventional						
DW	25.4	29.0	2.5	6.3	690.0	69.0
FW	31.0	35.33	2.6	4.3	625.0	62.5
UTL	31.0	35.33	2.6	4.3	625.0	62.5
SEm (±)	0.54	0.63	0.04	0.09	9.57	0.9
CD (0.05%)	1.88	2.20	0.15	0.32	33.12	3.31

Values are given in average of three replicates

Table 8.3: Effect of jowar grain spawn on the growth and yield of *P. sajor-caju* on conventional and non-conventional substrates

Substrates	Spawn run (day)	Pin head (day)	Stipe length (cm)	Cap diameter (cm)	Total yield (gm/kg)	BE (%)
Conventional						
SS	19.0	22.7	2.7	7.7	870.0	87.0
WS	21.0	25.0	2.6	8.5	760.0	76.0
PS	22.7	27.0	2.8	0.43	710.0	71.0
SEm (±)	0.50	1.24	0.18	0.12	18.70	1.87
CD (0.05%)	1.76	4.31	0.63	0.43	64.73	6.47
Non-conventional						
DW	26.7	30.7	2.5	6.2	680.0	68.0
FW	31.0	35.4	2.6	5.0	616.7	61.6
UTL	27.0	31.0	2.4	5.6	606.7	60.6
SEm (±)	1.24	1.56	0.12	0.18	15.12	1.51
CD (0.05%)	4.31	5.40	0.43	0.63	52.33	5.23

Values are given in average of three replicates

Discussion

Our present study clearly indicate that wheat grains spawn are best for cultivation of *Pleurotus sajor-caju* followed by sorghum grain spawn. Maize grain spawn was comparatively less effective for cultivation of *Pleurotus sajor-caju*. Selection of a suitable strain is the prime and important aspect in the spawn preparation. Second aspect of a quality spawn is selection of a suitable spawn substrate. Even lot of spawn substrates have been suggested by various workers, grain spawn is mostly suited for oyster mushrooms. Commonly used grains are wheat, sorghum, maize, jowar and paddy. The choice of grain depends on the availability of the same in a particular locality. Variations in spawn run rate and yield may be attributed to the size of the grains. Smaller grains have a greater number of inoculation points per kg than larger grains (Mamiro and Royse, 2008). It was found that the spawn run rate of smaller grains was higher than the larger grains. However, larger grains have a greater food reserve (Elliot, 1985) and can sustain the mycelium for longer periods of time during stress (Fritsche, 1988). Thus, different types of spawn may influence productivity and growth. These observations are in agreement with the result of Jain, 2005 who also found wheat grains are more suitable for cultivation of *Pleurotus* sp.

Many workers worked on development of different grain spawns and their effect on yield (Elliot, 1985; Fritsche, 1988; Sharma, 2003; Chaurasia, 1997; Royse, 2002; Jain, 2005; Shah *et al.*, 2004; Arulnandhy and Gayathri., 2007; Pathmasini *et al.*, 2008). Pathmasini *et al.*, 2008 used locally available grains of kurakkan (*Eleusine coracana*), maize (broken), sorghum (*Sorghum bicolor*) and paddy (*Oryza sativa*) for spawn production. According to him the kurakkan spawn produced an acceleration of spawn running, pin head formation, fruit body formation and increased yield compared with other types of spawn *viz.* maize, sorghum and paddy. Shah *et al.*, (2004) and Tan (1981) took three types of grain for spawn production kurakkan (*Eleusine coracana*), maize (broken) (*Zea mays*), sorghum (*Sorghum bicolor*) and reported the spawn run appear earlier in kurukkan. Arulnandhy and Gayathri (2007) obtained a mean yield of 24 g on sawdust medium. Chaubey, (2010) used wheat, maize and sorghum grain spawn for the cultivation of oyster mushroom. Thulasi *et al*, (2010) reported spawn production of oyster mushroom on different substrates. Khan *et al*, (2011) reported different spawning methods of oyster mushrooms on cotton waste. It is concluded from the present study that all the three grains are suitable for spawn production but wheat grain is more efficient for the spawn production of *Pleurotus sajor-caju.*

References

1. Arulnandhy, V. and Gayathri., T.(2007). Identification of suitable and efficient substrate for the production of oyster (*Pleurotus ostreatus*) mushrooms. Undergraduate research report, Department of Agricultural Biology, Eastern University, Sri Lanka.

2. Bano, Z. and Rajarathanam, S. (1982). Studies on the cultivation of *Pleurotus sajor-caju, The Mushroom Journal.* 115: 243-245.

3. Chaubey, A. (2010). Studies on cultivation technology of medicinal mushrooms with special reference to marketing potential in Bundelkhand region. Ph. D. Thesis. Dr. H.S. Gour, University, Sagar, M.P.

4. Chaubey, A., Dehariya, P. and Vyas, D. (2010). Solid waste management through oyster mushroom cultivation. *International J. of Biozone.* 2 (1&2).369-372.

5. Chaurasia, V.K. (1997). Studies on production technology of *Pleurotus columbines* at Raipur. M.Sc. Thesis submitted at I.G.A.U. Raipur (C.G.).

6. Dehariya P and Vyas D (2013). Effect of different agro-waste substrates and their combinations on the yield and biological efficiency of *Pleurotus sajor- caju. IOSR J of Pharmacy and Biological Sciences* 8 (03): 60-64.

7. Dehariya P, Chaubey A, Vyas D (2010). *Pleurotus sajor-caju*: an alternative source for bioconversion of wastes. *Mushroom Research* 19: 90-93.

8. Elliot, T.J. (1985). Spawn – making and Spawns. In: *The Biology and Technology of the Cultivated Mushrooms,* P.B. Flegg, D.M. Spencer and D.A. Wood (Eds.), John Wiley & Sons Ltd. pp. 131-139.

9. Fritsche, G. (1988). Spawn: properties and preparation, In: *The Cultivation of Mushrooms.* van Griensven, L.J.L.D. (Eds.), Darlington Mushroom Laboratories, Sussex. pp. 1–99.

10. Jain, A. K. (2005). Thesis on Mushroom Cultivation with special reference to *Pleurotus florida* and their Marketing potential in Sagar Region.

11. Jain, A.K. and Vyas, D. (2005). Comparative study on the yield of three *Pleurotus* sp. grown in several lignocelluloses By- products. *J. Basic Appl. Mycol.* 4: 155-157.

12. Jain, A. K. and Vyas, D. (2002). Yield response of *Pleurotus florida* on wheat straw in combination with other substrate. *Mush. Res.*11: 19-20.

13. Jain AK. and Vyas D. (2005). Supplementation of Soybean choker: Enhances the growth and yield of *P. sajor- caju* grown in lignocellulosic waste. *J. Basic and Appl. Mycol* 3 & 4: 88-90.

14. Khan, M.W. Ali, M, A, and Farooq, M. (2011). Growth respons of oyster mushroos (*Pleurotus ostreatus*) by using different methods of spawning raised on cotton waste. *Pak. J. Phytopathol.,* 23 (2): 156-158.

15. Mamiro, D.P. and Royse, D.J. (2008).The influence of spawn type and strain on yield, size and mushroom solids content of *Agaricus bisporus* produced on non-composted and spent mushroom compost. *Bioresourse Technology.* 99: 3205-3212.

16. Madhusudhan, K. and Chandra Mohan, R. (2002). Effect of certain factors on the yield of *Pleurotus sajor-caju* on Areca wastes. *Indian Phytopathology.* 55(3): 371.

17. Pathak, V.N., Yadav, N. and Gour, M. (2000). Mushroom Production and Processing Technology. *Agrobios*, India.

18. Pathmashini, L., Arulnandhy, V. and Wilson, R.S. (2008). Cultivation of oyster mushroom (*Pleurotus ostreatus*) on saw dust. *Cey. J. Sci. (Bio. Sci.)* 37 (2): 177-182.

19. Quimio, T.H. (1978). Introducing *Pleurotus flabellatus* for your dinner table. *Mushroom Journal*, 69: 282-283.

20. Royse, D.J. (2002). Influence of spawn rate and commercial delayed release nutrient levels on *Pleurotus cornucopiae* (oyster mushroom) yield, size and time production. *Applied Microbial and Biotechnol.* 58: 527-531.

21. Royse, D.J. and Schisler, L.C. (1987a). Yield and size of *Pleurotus ostreatus* and *Pleurotus sajor-caju* as effected by delayed release nutrient supplementation. *Applied Microbial* and *Biotechnol.* 26: 191-194.

22. Royse, D.J. and Schisler, L.C. (1987b). Influence of benomyl on yield response of *Pleurotus sajor-caju* to delayed release nutrient supplementation. *Horticulture Science.* 22: 60-62.

23. Shah, Z.A., Asar, M. and Ishtiaq (2004). Comparative study on cultivation and yield performance of oyster mushroom on different substrates (wheat straw, leaves, saw dust). *Pakistan J. Nutrition* 3: 159-160.

24. Sharma, B.B. (2003). Effect of different substrates on spawn growth and yield of pink oyster mushroom *Pleurotus djamor. J. of Mycol. and Pl. Pathol.* 33 (2): 265-268.

25. Tan, T.T. (1981). Cotton waste is a good substrate for cultivation of *Pleurotus ostreatus* the oyster mushroom. *Mushroom Science.*11: 705-710.

26. Thulasi, E. P. Thomas, D., Ravichandran, B. and Madhusudhanan, K. (2010). Mycelial culture and spawn production of two oyster Mushrooms, *Pleurotus florida* and *Pleurotus eous* on Different Substrates. *International Journal of Biological Technology.* 1(3): 39-42.

Sustainable Development of Natural Resources and
Wildlife Conservation
Editor: **Ashwani Kumar Dubey**
Published by: **REGENCY PUBLICATIONS, NEW DELHI**

9

Physico-chemical Analysis and Chemo Profiling of Ayurvedic Single Drugs of Herbal Origin–*Santalum album* Linn.

Deepak Mishra #, Pratima Mishra* and Arpita Awasthi **

ABSTRACT

Herbal drugs are traditional medicine based on the use of plants and plant extracts. The use of herbal supplements has increased dramatically over the past 30 years. It is getting popularized in developing and developed countries owing to its natural origin and lesser side effects. In present study Ayurvedic formulated drug Santalum album Linn. was subjected to physico-chemical analysis and chemo profiling to fix the quality standards of this drug. It is one of the important herbal plant mentioned in Ayurveda. It is cultivated for its aromatic wood and oil. It is widely distributed in the Indo-Malesian region and Peninsular India. Sandalwood is used for the treatment of acute dermatitis, bronchitis, cystitis, eye diseases, gonorrhea, herpes, zoster, infection, palpitations, sunstroke, urethritis, vaginitis, psychopathic, Skin disorders, Heart ailments, Anti-pyretic, General weakness, Urinary tract infection and many more. During physico-chemical analysis and chemo profiling churna was prepared as siddhyog sangrah and the preparation are standardized according to guidelines of Ayurvedic Pharmacopoiea, viz extractive value, Ash value, pH value,

#Department of Biotechnology, A.K.S. University Satna (M.P.)
*Department of Biotechnology, Pentium Point College Rewa (M.P.)
**Department of Botany and Microbiology, T.R.S. College, Rewa (M.P.)
Corresponding author: E-mail: deepakrewabiotech@gmail.com

Moisture contents etc. Various chromatographic techniques have been carried out for chemo profiling of drug. These studies on Santalum album (wood) were reproducible, precise and may be considered as a method for its quality control.

Key words: *Standardization, Ayurvedic formulation.*

Introduction

Ayurveda is one of the great gifts of the sages of ancient India to mankind. It is one of the oldest scientific medical systems in the world, with a long record of clinical experience. However, it is not only a system of medicine in the conventional sense of curing disease. It is also a way of life that teaches us how to maintain and protect health[1]. It shows us both how to cure disease and how to promote longevity. Ayurveda treats man as a "whole" – which is a combination of body, mind and soul. Therefore it is a truly holistic and integral medical system.

In Ayurveda herbal medicines are basically used as a part of samana chikitsa. These are traditional remedies based either on whole plant or plant extracts. In the last few decades there has been an exponential growth in the field of herbal medicine. It is getting popularized in developing and developed countries owing to its natural origin and lesser side effects[2]. In olden times, *vaidyas* used to treat patients on individual basis, and prepare drug according to the requirement of the patient. But the scene has changed now; herbal medicines are being manufactured on a large scale in mechanical units, where manufacturers own across many problems such as availability of good quality raw material, authentication of raw material, availability of standards, proper standardization methodology of single drugs and formulations, quality control parameters, etc[3].

Santalum album Linn. is one of the important herbal plant used in ayurveda for the treatment of various diseases. It is member of family Santalaceae and commonly known as sweta chandan. It is widely distributed in throughout the India especially in Indo-Malesian region and in the dry regions of peninsular India[3]. Though it is naturalized in many parts of India *i.e.* in Vindhya Mountains southwards, also in Karnataka, Andhra Pradesh and Tamil Nadu; it is cultivated for its aromatic wood and oil[4]. Sweta chandan is a small to medium sized, evergreen semi-parasitic tree, with slender branches, sometimes reaching up to 18 m in height and 2.4 m in girth. Barth reddish or dark grey or nearly black, rough with deep cracks on old trees; leaves glabrous, thin, elliptic-ovate or ovate-lanceolate, 1.5-8 cm* 1.6-3.2 cm, sometimes larger; flowers straw-coloured, brownish purple, reddish purpleor violet[5]. Sandal is capable of growing in different kinds of soil like sand, clay, laterite, loam, black-cotton etc[6, 7]. It is capable of regenerating profusely in the absence of fire and grazing[8]. Sandalwood is used for acute dermatitis, bronchitis, cystitis, eye diseases, gonorrhea, herpes, zoster, infection, palpitations, sunstroke, urethritis, vaginitis, psychopathic, Skin disorders, Heart ailments, Anti-pyretic, General weakness, Urinary tract infection and many more[9, 10].

Materials and Methods

Selection of plant material for study

In present work *Santalum album* **Linn.** (Swetachandan) have been selected for the study. It has been collected from Chitahara forest (Manjhagawa) of Satna district. The plant has identified on the basis of different pharmacopeial and botanical standard. Mostly extract of wood has been used in the study[11].

Physico-chemical Profiling

Determination of Moisture Content "Loss on Drying" (LOD)

Place about 2gm of drug (without preliminary drying) after accurately weighing (accurately weighed to within 0.01g) it in a tarred evaporating dish. After placing the above said amount of the drug in the tarred evaporating dish dry at 105°C for 5 hours, and weigh. Continue the drying and weighing at one hour interval until difference between two successive weighing corresponds to not more than 0.25 per cent. Constant weight is reached when two consecutive weighing after drying for 30 minutes and cooling for 30 minutes in a desiccator, show not more than 0.01g difference[12].

Determination of Water and Alcohol Soluble Extractive

2gm of the air dried drug, coarsely powdered with 100ml of water and alcohol respectively of the specified strength in a closed flask for twenty-four hours, shaking frequently during six hours and allow it to stand for eighteen hours. Filter rapidly and evaporate 25ml of the filtrate to dryness in a tarred flat bottomed shallow dish, and dry at 105°C, to constant weight and weigh.

Determination of Total Ash Value

2 to 3gm of drug has been incinerated in a tarred platinum or silica dish at a temperature not exceeding 450°C until free from carbon, cool and weigh. More specifically carbon free ash can be obtained by exhaust the charred mass with hot water, collect the residue on an ash less filter paper, incinerate the residue and filter paper, add the filtrate, evaporate to dryness and ignite at a temperature not exceeding 450°C. Calculate the percentage of ash with reference to the air-dried drug.

Determination of Acid Insoluble Ash

Boil the ash obtained in above process for 5 minutes with 25 ml of dilute hydrochloric acid ; collect the insoluble matter in a gooch crucible, or on an ashless filter paper, wash with hot water and ignite to constant weight calculate the percentage of acid-insoluble ash with reference to air dried drug.

Determination of Water Soluble Ash

Boil the ash obtained in total ash value for 5minutes with 25 ml of distilled water; collect the insoluble matter in a gooch crucible, or on an ash less filter paper, wash with hot water and ignite to constant weight calculate the percentage of water-soluble ash with reference to air dried drug.

Chemo profiling of Drug

HPTLC-extract the crude samples in alcohol (150 ml×5), concentrate to 5-10 ml on water bath and carry out High performance thin layer chromatography by applying 4µl of the sample on pre-treated TLC silica gel plate 60 F254 (from Merck India Ltd, Germany) and develop the plate to a distance of 10 cm using Toluene: Ethyl acetate: Formic acid (7:2.5:0.5) as mobile phase. After derivatization allow the plates to dry in room temperature & examine under ultra violet light at 254 nm; under366 nm; after derivatization with methanol-sulphuric acid reagent. Measure and record the distance of each spot from the point of its application and calculate the fro value by dividing the distance traveled by the spot by the distance traveled by the front of the mobile phase.

Results and Discussion

Standardization of drugs and formulation of traditional system of medicine is very important for assuring uniform quality and optimum biological efficiency[13]. Standardization of *Santalum album* Linn. becomes very helpful in establishing standards for quality control analysis. The physico-chemical parameters identified represent the specific diagnostic characters of drug. This study facilitates in identifying the adulterant or substitutes of the ingredients used in the formulation. Chemo profiling by HPTLC has helped in developing identification tests and finger printing of constituents of the ingredients used in the formulations.

During physico chemical analysis 5.50% Moisture Content "Loss on Drying" (LOD) has been observed which should be not less than 5.0% (table 9.2). In the analysis of water soluble extractive value should not be more than 12% which is 4.0% has been observed in sweta chandan (table 9.3). Along with this 2.0% alcohol soluble extractive value has been detected which should be less than 5.0% (table 9.4). 4.5% total ash value has been observed which is also within the standard limit which is 5% (table 9.5). Generally acid insoluble ash value should be less than 2.0% which has 3.0%

has been observed (table 9.6). Also water soluble ash observed is 3.5% has been observed which should be less than 2%.

In association of these physico chemical studies also chemo profiling studies has been conducted. In which we study about comparative profile of wild drug (A) in respect to market sample (B):

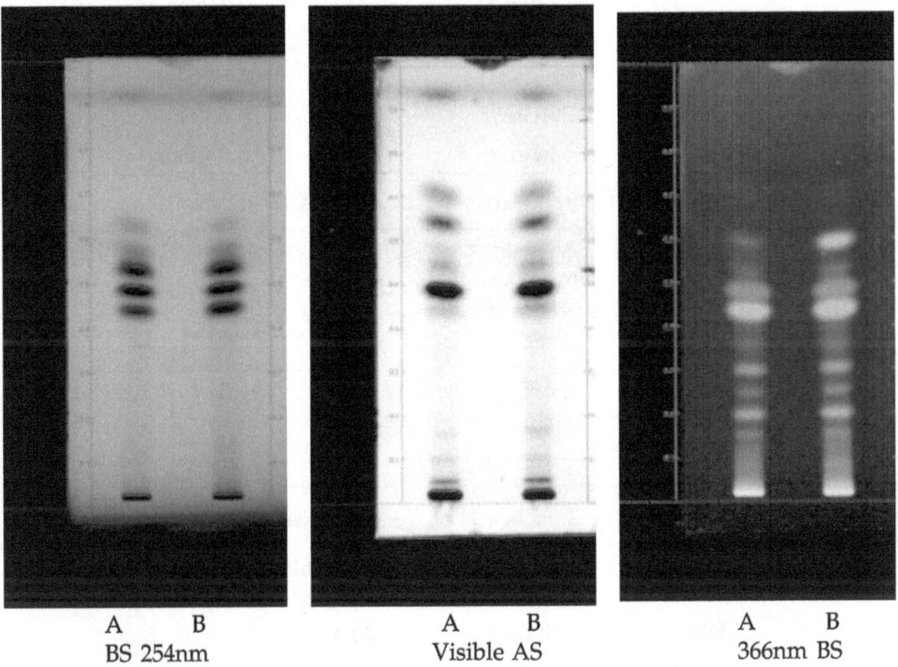

Fig. 9.1. Chromatographic Figure Print of Sweta Chandana

Table 9.1: R_f value in Test Solution of *Santalum album* Linn.

R_f - value	Before derivatization			
	Rf at 254 nm		Rf at 366 nm	
	Sample A	Sample B	Sample A	Sample B
R_f -1	0.43(Black)	0.43(Black)	0.05(blue)	0.05(blue)
R_f -2	0.48(Black)	0.48(Black)	0.11(Green)	0.11(Green)
R_f -3	0.52(Black)	0.52(Black)	0.15(Green)	0.15(Green)
R_f -4	0.56(Gray)	0.56(Gray)	0.20(Green	0.20(Green
R_f -5	0.63(Gray)	0.63(Gray)	0.24(blue)	0.24(blue)
R_f -6	-	-	0.30(sky blue)	0.30(sky blue)
R_f -7	-	-	0.44(blue)	0.44(blue)
R_f -8	-	-	0.48(blue)	0.48(blue)
R_f -9	-	-	0.52(green)	0.52(green)
R_f -10	-	-	0.60(Green)	0.60(Green)

contd...

Table 9.1: *contd...*

R_f - value	After derivatization Rf at Visible light	
	Sample A	**Sample B**
R_f -1	0.43(Black)	0.43(Black)
R_f -2	0.48(Black)	0.48(Black)
R_f -3	0.52(Black)	0.52(Black)
R_f -4	0.56(Grey)	0.56(Grey)
R_f -5	0.63(Grey)	0.63(Gray)

Table 9.2: Moisture Content "Loss on Drying" (LOD) (Sweta chandan)

S.No.	Petridish empty weight	Preweight emp.Wt.+2g drug (A)	Post weight		Difference (A-B)
			1st	2nd (B)	
1.	29.65	31.65	31.55	31.54	0.11
2.	33.38	35.28	35.18	35.17	0.11
3.	26.30	28.30	28.18	28.18	0.12

Table 9.3: Water Soluble Extractive value (Sweta chandan)

S.No.	Water sample	Pre weight (A)	Post weight (B)	Difference (B-A)
1.	MW1	43.06	43.10	0.05
2.	MW2	26.29	26.32	0.03
3.	MW3	29.64	29.68	0.04

Extractive value % = Avg. X 100 = 0.04× 100 = 4.0%

Table 9.4: Alcohol Soluble Extractive value (Sweta chandan)

S.No.	Alcohol sample	Pre weight (A)	Post weight (B)	Difference (B-A)
1.	MA1	26.30	26.32	0.02
2.	MA2	40.73	40.75	0.02
3.	MA3	48.53	48.54	0.01

Extractive value % = Avg.X 500 = 0.02× 100 = 2.0%

Table 9.5: Total Ash Value (Sweta chandan)

S.No.	Crusible weight (A)	1st	2nd	3rd (B)	Difference (B-A)
A	39.17	39.32	39.29	39.29	0.12
B	36.17	33.25	36.22	36.22	0.05
C	37.03	36.17	36.14	36.14	0.11

Ash value = Average× 50 = 0.09× 50 = 4.5 %

Table 9.6: Acid Insoluble Ash (Sweta chandan)

S.No.	Crusible weight(A)	1st	2nd (B)	3rd	Difference (A-B)
1	39.29	39.28	39.26	39.26	0.03
2	40.82	40.80	40.78	40.78	0.04
3	33.59	33.50	33.48	33.48	0.11

Ash value   =   Average× 50 =   0.06× 50   =   3.0 %

Table 9.7: Water Soluble Ash (Sweta chandan)

S.No.	Last weight of Total ash (A)	1st	2nd (B)	3rd	Difference (A-B)
1	40.82	40.80	40.78	40.78	0.04
2	33.59	33.50	33.48	33.48	0.11
3	36.22	36.20	36.17	36.17	0.05

Ash value   =   Average× 50 =   0.07× 50   =   3.5 %

Conclusion

The acceptability and utility of herbal formulations is many a time eclipsed for lack of acceptable norms of quality control parameters[14]. These quality parameters / analytical values developed in the present study can be utilized by in-house quality control laboratory of an Ayurvedic pharmaceutical unit to evaluate the genuineness and authenticity of plant material available.

From overall data it is concluded that Herbal drug standardization is massively wide and deep[15]. Quality control of herbal medicines has not only to establish reasonable analytical methods for analyzing the active constituents in herbal medicines, but many other factors should be concerned, such as pesticides residue[16], aflatoxine content[17], the heavy metals contamination, good agricultural practice (GAP), good manufacturing practice (GMP), etc[18]. There is so much to know and so much seemingly contradictory theories on the subject of herbal medicines among the existing analytical methods, chromatographic methods are still the mainstream, due to their integrative evaluation characteristics[19].

References

1. Caius J F., (1990). The Medicinal and Poisonous Plants of India *Indian Medicinal Plants*; 2:489.

2. Devi M Vimala, (2000). Quality Control and Assurance of Indian Medicines *Health Administrator* Vol: XX Number 1&2 : 21-25 Pg.

3. Baldovini N., Delasalle C., and Joulain D., (2011). Phytochemistry of the heartwood from fragrant Santalum species: a review *Flavour and Fragrance Journal*; 26:7-26.

4. Samantray Sanghamitra, Upadhyaya Chandni (2010). Methodological studies and research on micropropagation of Chandan (*Santalum album L.*): An endangered plant *International Journal on Science and Technology (IJSAT)* Volume 1, Issue I, 010-018.

5. Banerjee S., Ecavade A., and Rao A.R. (1993). Modulatory influence of sandalwood oil on mouse hepatic glutathione S-transferase activity and acid soluble sulfhydryl level *Cancer Letters*; 68: 105-109.

6. Jain S.H., Angadi V.G. and Shankaranarayana K.H. (2003). Edaphic, Environmental and Genetic Factors associated with Growth and adaptability of Sandal (Santalum album L.). *In provenances. Sandalwood Research Newsletter*:17, 6-7.

7. Anjum Perveen And Mohammad Qaiser (2005). Pollen Flora of Pakistan–XLIII Lythraceae *Pak. J. Bot.*, 37(1): 1-6.

8. Fox, J.E. (2000). Sandalwood: the royal tree *Biologist (London)*; 47:31-34.

9. Balachandran P, Govindrajan R. (2005). Cancer-An ayurvedic perspective *Pharmacol Res*; 51: 19-30.

10. Srinivasan, V.V., Sivaramakrishnan, V.R., Rangaswamy, C.R., Anantha padmanabha, H.S. and Shankaranarayana, K.H. (1992). *Sandal (Santalum album L.)*, ICFRE, Dehra Dun.

11. M. Rajani and Niranjan S. Kanaki (2008). Phytochemical Standardization of Herbal Drugs and Polyherbal Formulation *Bioactive Molecules and Medicinal Plants*, 10, 349-369.

12. Blois M.S. (1958). Antioxidant determinations by the use of a stable free radical *Nature*; 181:1199-1200.

13. Brahmachari U.N. (2001). The role of science in the recent progress of medicine *Current Science*, Vol. 81, No. 1.

14. Jigna Parekh, Sumitra V. Chanda (2008). Antibacterial Activity of Aqueous and Alcoholic Extracts of 34 Indian Medicinal Plants against Some Staphylococcus Species *Turk J Biol* 32, 63-71.

15. Singh Himmat, Mishra Swatantra Kumar and Pande Milind (2010). Standardization of *Arjunarishta* Formulation By TLC *Method International Journal of Pharmaceutical Sciences Review and Research*; 2(1), 25-28.

16. Battaglia S. (2003). The Complete Guide to Aromatherapy *The International Centre of Holistic Aromatherapy*, Brisbane, Australia.

17. Tambekar D.H. and Dahikar S.B. (2010). Exploring antibacterial potential of some ayurvedic preparations to control bacterial enteric infections *Journal of Chemical* and *Pharmaceutical Research*; 2(5): 494-501.

18. Singh Rana Pratap and Singh D.A. (2011). Antibacterial activity of Alcoholic and Aqueous extracts of some Medicinal Plants. *International Journal of PharmTech Research*; 3(2), 1103-1106.

19. Patra Kartik Chandra, Pareta Surendra K., Harwansh Ranjit K., Kumar K. Jayaram (2010). Traditional Approaches towards Standardization of Herbal Medicines-A Review, *Journal of Pharmaceutical Science and Technology*, Vol. 2 (11), 372-379.

Sustainable Development of Natural Resources and
Wildlife Conservation
Editor: **Ashwani Kumar Dubey**
Published by: **REGENCY PUBLICATIONS, NEW DELHI**

10

Preliminary Study of Noise Pollution in Kota City, Rajasthan, India

Prahlad Dube

ABSTRACT

Noise means any unwanted sound. Sounds particularly loud ones that disturb people or make it difficult to hear wanted sounds are noise. Noise pollution is the disturbing or excessive noise that may harm the activity or balance of human or animal life. The source of most outdoor noise worldwide is mainly caused by machines and transportation systems, motor vehicles, aircraft, and trains. Poor urban planning may give rise to noise pollution, since side-by-side industrial and residential buildings can result in noise pollution in the residential areas. Noise pollution affects both health and behaviour. Keeping this point in mind, present study was taken up. For the samples of noise level, ten (10) points were randomly selected where local residents complaint high nose level during an earlier survey carried out by my students. The sound meter was used to record the noise level. The readings were taken thrice a day (6-7 AM, 4-5 PM and 8-9 PM). The readings were compiled, dealt statistically and inference was drawn. The results of present preliminary investigation shows that the noise level has reached an alarming level in 70% study points and it was found to be above the standard level of Central Pollution Control Board. Highest noise level points were station area, Kotari circle, Mahaveer Nagar area and lowest noise level points were Ganesh Nagar and University of Kota areas. The present study indicates

Biodiversity Research Unit, Department of Zoology, Government College, Kota: 324009, Rajasthan, India.
Corresponding author: Email: dube_prahlad@rediffmail.com

an urgent need of a systematic, wider and long term multidimensional study in this field.

Key words: *Noise, noise pollution, highest noise level, health and behaviour.*

Introduction

NOISE refers to "Noise means any unwanted sound." It include: - Sounds particularly LOUD ONES that disturb people or make it difficult to hear wanted sounds are noise. Noise pollution is the disturbing or excessive noise that may harm the activity or balance of human or animal life. The source of most outdoor noise worldwide is mainly that caused by machines and transportation systems, motor vehicles, aircraft and trains. Poor urban planning may give rise to noise pollution, industrial and residential buildings can result in noise pollution in the residential areas. Noise pollution affects both health (hearing impairment, hypertension, ischemic heart disease, annoyance, and sleep disturbance, changes in the immune system and birth defects) and behavior (stress, restlessness, anger, depression and psychological problems).

Study Area

Rajasthan known as "the land of kings" and is the largest state of the Republic of India by area. It is located in the west of India. It comprises most of the area of the large, inhospitable Thar Desert. The first mention of the place name Rajasthan appears in James Tod's 1829 publication, Annals and Antiquities of Rajasthan or the Central and Western Rajpoot States of India. Rajasthan literally means a Land of Kingdoms. George Thomas (Military Memories) was the first in 1800 A.D., to term this region as Rajputana. Geographically, the state is divided by The Aravalies (Oldest hill chain) into two distinct Zones: 1. Western: arid and semi-arid zone 2. Eastern: fertile and water rich zone. Kota city is situated in south-eastern part of Rajasthan. It is an industrial city and famous as education hub. It's population is about 15 lacs and about 2 lacs students come every year here to get coaching for various competitive exams (important of them are JEE, AIPMT etc.). Therefore, a regular human activity – both in residential and commercial areas continued most of the day. A preliminary survey was conducted through under graduate students of my department to know the effects of coaching students' activity on noise levels of city which made the basis of present work.

Methodology

For the samples of noise level, ten (10) points were randomly selected where local residents complained high nose level during an earlier survey carried out by my students. All sites (points) were located within the periphery of 25 km. The SOUND METER was used to record the noise level. The readings were taken Thrice a Day (6-7 AM, 4-5 PM and 8-9 PM). The study was carried out during May — July 2013.The readings were compiled, dealt statistically and inference was drawn.

Results and Discussion

The observations recorded during the present study are put in Table 1 and are represented by bar diagrams in Fig. 10.1. The results of present preliminary investigation show that the noise level has reached an ALARMING LEVEL in 70% study points and it was found to be above the standard level of Central Pollution Control Board. Highest noise level points were station area, Kotari circle, Mahaveer Nagar area and lowest noise level points were Ganesh Nagar and University of Kota area. Sites 03, 05 and 06 were all time high— noise level zones. This is causing health disorders and behavioural changes in the residents of the city. Although

Table 10.1: Readings of Noise level (dB) in various study sites at Kota (Rajasthan)

Time	Site 1	Site 2	Site 3	Site 4	Site 5	Site 6	Site 7	Site 8	Site 9	Site 10
6-7 AM	56.5	61.6	57.8	50	67.6	88	54.5	55.6	47	60
4-5 PM	58.2	58.9	65.4	57.8	69.9	95.6	58.8	59.9	51.9	68.9
8-9 PM	68.3	52.5	57.3	54.4	70.7	93.2	50.7	51.9	51.3	52.3

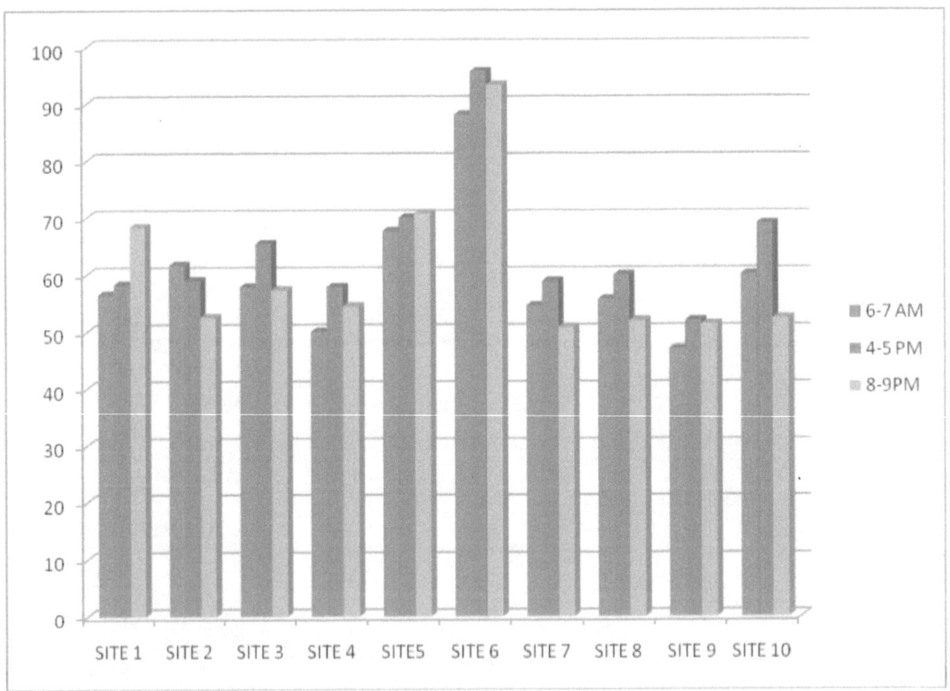

Fig. 10.1. Showing Noise level (dB) in various study sites at Kota (Rajasthan)

these data are not sufficient to draw any valid inference but in the light of earlier investigations carried out in other cities of India (Singh and Davar 2004, Pawar and Joshi 2005, Vidya Sagar and Rao 2006, Gangwar *et al.* 2006 and Sharma *et al.* 2010), it can be concluded that noise level is increasing and "high dB – Zones" are increasing in the city. A deep, systematic and long time study is needed to deal with this severe and alarming problem.

References

Gangwar, K. K., Joshi, B. D. and Swami, A. (2006) Noise pollution status at four selected intersections in commercial areas of Bareilly metropolitan city. *Him. J. Env. & Zool.*, 20(1): 75.

Pawar, C. T. and Joshi, M. V. (2005) Urban development and sound level in Ichalkaranji city, Maharashtra. *Indian J. Environ. & Ecoplan.*, 10(1): 177.

Sharma, V., Saini, P., Kaushik, S. and Joshi, B. D. (2010) Assesment of Noise level in different zones of Haridwar City of Uttarankahand State, India. New York Sci. J., 3(4):109 -111.

Singh, N. and Davar, S. C. (2004) Noise Pollution- Sources, Effects and Control. *J. Human Ecol.*, 16 (3): 181.

Vidya Sagar, T. and Nageshwara Rao, G. (2006) Noise Pollution Levels in Visakhapatnam City (India). *J. Env. Sci. and Engg.*, 48 (2): 139.

Sustainable Development of Natural Resources and
Wildlife Conservation
Editor: **Ashwani Kumar Dubey**
Published by: **REGENCY PUBLICATIONS, NEW DELHI**

Pages 73-77

11

Diversity of Phytoplankton in Kishore Sagar Tank, Kota, Rajasthan

Seema Sharma* and Prahlad Dube**

ABSTRACT

Hadoti region in Rajasthan which includes four Revenue districts viz; Kota, Bundi, Jhalawar and Baran, is rich in natural as well as manmade water resources. Kishor Sagar tank is situated in the heart of the Kota city. Community structure of phytoplankton in Kishore Sagar tank Kota (Rajasthan) was investigated for a period of two years. During the investigation 24 species of phytoplanktons of 17 families related to 5 phyla were recorded. The dominant species of phytoplanktons Spirogyra varians, Oscillatoria princeps, Nostoc species were recorded. The results obtained were discussed with reference to published literature. The study reveals that Kishore Sagar tank Kota is severely affected due to anthropogenic activities. The phytoplankton community indicates the pollution status of Kishore Sagar.

Key words: *Kota, kishore sagar tank, phytoplankton*

Introduction

Phytoplanktons are the base of aquatic food webs and energy production is linked to phytoplankton primary production. Excessive nutrient and organic inputs from human activities in lakes and their watersheds lead to

*Department of Bioscience & Biotechnology, Banasthali Vidyapeeth (Rajasthan)
**Biodiversity Research Lab, Department of Zoology, Government College, Kota (Rajasthan) India.
Corresponding author: Email- mission.seema21@gmail.com

eutrophication, characterized by increases in phytoplankton biomass, nuisance algal blooms, loss of water clarity from increased primary production and loss of oxygen in bottom waters. Phytoplankton is the base of food web which affects the food production. Diatoms have been used by ecologists to indicate pollution in water body and other variation of ecological conditions. Chlorophyta was dominating group and cyanophyta was second dominate group. In the tank the contribution of chlorophyta and cyanophyta to total biomass was higher than 90%. Phytoplankton biomass reached the value many times higher.

The value of plankton and other algae as direct or indirect food for fish and their usefulness as indicators of water conditions have long been well recognized. Phytoplanktons are the base of aquatic food webs and energy production is linked to phytoplankton primary production. Excessive nutrient and organic inputs from human activities in lakes and their watersheds lead to eutrophication, characterized by increases in phytoplankton biomass, nuisance algal blooms, loss of water clarity from increased primary production and loss of oxygen in bottom waters.

The earliest contribution known to the studies related to planktons were made by Allen (1920), Arora (1931), Beadles (1932), Alikunhi (1952), Welch (1952), Ruttner (1953), Rao *et al.* (1977) and Dad (1981).

The knowledge of plankton in Rajasthan is fragmentary though a number of contributions are available. Gautam (1982, 1989), Sharma and Saksena (1983), Choube (1987), Sharma (1991), Sharma and Sharma (1992), Verma (2001), Dube (2002), Durve and Rao (2006) are some of the contributors. Rao *et. al.* (2006) have studied concept of plankton species diversity in small water body Lake Rangasagar (Udaipur: Rajasthan). The qualitative zooplankton communities were examined for a period of two year. The results revealed that higher plankton diversity need not necessarily lead to a greater community diversity or evenness of species diversity.

Study Area

Present work is based on the two-year long limnological study made on Kishore Sagar Tank. Kishore Sagar Tank is situated in the heart of the Kota city. Maharaja Dhir Dev built this tank in the year 1346AD. Maharao Kishore Singh of Kota developed this tank, decorated it with beautiful sculpture around it, and renamed as Kishore Sagar. Area of the tank is 7.50 lakh sq. m. Normal capacity of the tank is 27,43,902 cubic m. and seasonal capacity is 56,47,865 cubic m., normal depth of the tank is 12-13 feet and seasonal depth is 23-24 feet. Rainfall of Kota district is 1500 mm and of Kota city is 900 mm. Kishore Sagar Tank is surrounded by road on which vehicular traffic runs round the clock.

Table 11.1: Qualitative estimation of phytoplankton in Kishore Sagar Tank

Phylum	Class	Family	*Genus and species*
Chlorophyta	Chlorophyceae	Hydrodictyaceae	*Hydrodictyon* sp.
Chlorophyta	Chlorophyceae	Chlamydomonadaceae	*Chlamydomonas eugametos*
Chlorophyta	Chlorophyceae	Volvocaceae	*Volvox globater*
Chlorophyta	Chlorophyceae	Oedogoniaceae	*Oedogonium nodulosum*
Chlorophyta	Chlorophyceae	Desmediaceae	*Closterium* sp.
Chlorophyta	Chlorophyceae	Zygnemaceae	*Zygnema* sp.
Chlorophyta	Chlorophyceae	Hydrodictyaceae	*Pediastrum duplex*
Chlorophyta	Chlorophyceae	Chaetophoraceae	*Draparnaldiopsis* sp.
Chlorophyta	Chlorophyceae	Chlorellaceae	*Chlorella vulgaris*
Chlorophyta	Chlorophyceae	Zygnemaceae	*Spirogyra karnalae*
Chlorophyta	Chlorophyceae	Zygnemaceae	*Spirogyra varians*
Chlorophyta	Chlorophyceae	Zygnemaceae	*Spirogyra jogensis*
Bacillariophyta	Bacillariophyceae	Melosiraceae	*Melosira varians*
Bacillariophyta	Bacillariophyceae	Pinnulariaceae	*Pinnularia viridis*
Cyanophyta	Cyanophyceae	Chroococcaceae	*Chroococcus turgidis*
Cyanophyta	Cyanophyceae	Oscillatoriaceae	*Oscillatoria princeps*
Cyanophyta	Cyanophyceae	Nostocaceae	*Nostoc muscoru*
Cyanophyta	Cyanophyceae	Scytonemataceae	*Scytonema simplex*
Cyanophyta	Cyanophyceae	Rivulariaceae	*Gloeotrichia indica*
Cyanophyta	Cyanophyceae	Microcystaceae	*Microcystis aeruginosa*
Cyanophyta	Cyanophyceae	Microcystaceae	*Microcystis flosaquae.*
Xanthophyta	Xanthophyceae	Botrydiaceae	*Botrydium granulatum*
Xanthophyta	Xanthophyceae	Vaucheriaceae	*Vaucheria geminata*
Euglenophyta	Euglenophyceae	Euglenoidae	*Euglena viridis*

Materials and Methods

Water and bottom mud samples were collected regularly twice in a month on the sampling day between 9 to 11 am. Samples were taken in washed polythene bottles of five-litre capacity. Standard methods were followed for collection of water samples for planktons (APHA, AWWA, WEF, 1989). Planktons were collected through a plankton silk net no. 30 and preserved in 5% formalin solution. Slide of plankton were made for taking microscopic observations. Planktons were identified using the works of Tonapi (1980), Adoni (1985) and APHA (1998).

Results and Discussion

The present study highlights good phytoplankton diversity in the Kishore Sagar Tank. 24 species of phytoplanktons of 17 families related to 5 phyla were recorded. High densities of planktons were recorded in summer and beginning of monsoon. The distribution and abundance of phytoplankton in the tank under the present investigation appeared to be related with the favorable water pH, alkalinity and dissolved oxygen content of the tank. The high temperature on the surface water acts as an important barrier for the upward migration of phytoplanktons during noon. The surface layer of water was rich in plankton because of availability of food in this layer. In the present study phytoplanktons were always found in abundance in bottom layer of water at morning and evening while at noon, they were sometime absent at surface layer and found at bottom layer. So, it was observed that temperature was the most important factor for the vertical migration. Our observations revealed that Light, dissolved oxygen and free CO_2 also affect planktonic movement and is in agreement with earlier works (Beeton1960, Tash and Armitage 1960, Michael 1970 and Dube 2002). A large number of plankton indicates high eutrophic nature of the water body.

References

Dad N.K. (1981). Limnological studies on Chambal Allen W.E. 1920. A quantitative and statistical study of plankton of San Joaquin River and its tributaries near Stockton, in 1913. Uni. Calif. Publ. Zool. 22 : 1-292.

Arora G.L. (1931). Fauna of Lahore. 2 Entomostraca of Lahore. Bull. Dept. of Zool. Punjab Univ. 1: 62-100.

Alikunhi K.H. (1952). On the food of young carp fry. J. Zool. Soc. India 4: 77-84.

Adoni A.D.S. (1985). Workbook on limnology. Ban dana printing service, New Delhi.

APHA (1998). Standards method for the examination of water and wastewater, 20th edition APHA, AWWA, WEF, Washington DC.

Beadles L.C. (1932). Scientific results of the Cambridge expedition to the east African lake in relation to their Fauna and Flora. *Trans. Amer. Fish. Soc.* 157-211.

Beeton, A.M., (1960). The vertical migration of *Mysis relicta* in lakes Huron and Michigan. J. Fish.Res. Bd. Canada, 17: 517-539.

Boyd, C.E. (1981). "Water Quality in Warm Water Fish Ponds", Craftmaster Printers, Inc. Opelika, Alabama.

Battish, S.K. (1992). "Fresh water zooplanktons Of India", Oxford and IBH Publishing Co. Ltd. New Delhi.

Dube P. (2002). Ecobiology of seasonal water bodies in southeastern plateau of Rajasthan with special reference to amphibians. *Indian J. Eniron. Sc.* 6(2) pp. 135-139.

Durve V.S., Rao P.S. (2006). Variation in primary production and its interrelationship with chlorophyll in the lake of Jaisamand (Rajasthan) *J. Acta Hyocrimicha. Hydrobiologia* 15(4): 379-387.

Gautam P.C. (1982). Plankton ecology of Hadoti lakes.Acta Ecol 4(1): 9-22.

Gautam P.C. (1989). Plankton ecology of Jait Sagar lake (Bundi). *Acta Ecol.* 11 (1): 17-21.

Michael, R.G., (1970). Diurnal variations in physicochemical factors and zooplankton in the surface layers of three freshwater fish ponds. *Indian J. Fish.*, 13: 48-82.

Ruttner F. (1953). Fundamental of limnology Univ. Tronto Press 1-242.

Rao K.S., Shrivastava P. and Divan A.P. (1977). Limnological studies on a semitropical lake, Physical factors effecting *Lophopodella carteri* (Ectoprocts, Phylactolaemata) *Bioresearch* ix 2p. 67-70.

Rao, K.S. (1993). "Recent advancement in Fresh water biology", Anmol Publishing Pvt. Ltd. New Delhi.

Sharma S.P. and Saksena D.N. (1983). Seasonal variation of the copepode component of zooplankton in a perennial tank of Janaktal Gwalior (India). *Acta Hydrochim. Et hydrobiol* 11(4): 479-484

Sharma K.C. (1991). Algae of Anasagar Lake of Ajmer Rajasthan Geobios new report 10: 67-68.

Sharma Renu and Sharma K.C. (1992). Diatoms of Anasagar lake of Ajmer. *Acta Ecol.* 14(1).

Sustainable Development of Natural Resources and   *Pages 78-87*
Wildlife Conservation
Editor: Ashwani Kumar Dubey
Published by: REGENCY PUBLICATIONS, NEW DELHI

12

Wise Use of Wetland for Sustainable Livelihood through Participatory, Alternative and Ecofriendly Approach in Coastal Belts of Odisha

Bibhu Santosh Behera,* Anama Charan Behera, S.K.Rout, B.P. Mohapatra, B. Parasar, A.P. Kanungo and B.K. Mohanty**

ABSTRACT

Adaptive capacities in vulnerable wetlands are relate to coping mechanism at the grassroots level for sustainable livelihood and climate specific. Given the fact that the absolute number of hardcore people is increasing and a large part of them live in the low-lying areas, the challenge lies in capacity building of these vulnerable communities to cope with climate change impacts.

This paper focuses on Soil less agriculture (Hydroponics) as an alternative source of livelihood means for the communities having no lands for cultivation. Approximately half of Bangladesh is cover with wetlands. The prospect of massive and enormous productivity lies in the development of wetland resources. Bangladesh has the highest wetlands to total land ratio in the world.

The soil-less agriculture is an indigenous practice in the central southwestern part of Bangladesh. The people living within the wetland ecosystem utilize locally

*College of Agriculture, Department of Extension Education (OUAT), Bhubaneswar, Odisha, India
**Department of Economics, D.B. College, Turumunga, Keonjhar, Odisha
Extension Education Department, OUAT, Bhubaneswar, Odisha
Corresponding author: Email: behera.bibhusantosh38@gmail.com

available paddy straws, water hyacinths and various aquatic invasive plants for making the floating mat or organic bed on which crops, vegetables and seedlings are grown. The productivity of this farming system is much higher than that of terrestrials agricultural and supportive to open water fisheries. The compost manure generates from refused organic bed is nutrient enriched and acts as soil conditioner. It would be a major source of nutrients in aquaculture as well.

This farming system is capable to ensure more agriculture production by restoring wetlands from aquatic invasive plant. More over the technology is friendly to the ecology and ecosystem of wetlands.

Key words: *Wetlands, Soil less-Agriculture, Aquatic Invasive plant, Climate change*

Introduction

Hydroponics (a Greek word hydro means water and ponos, labour) a soil-less agriculture is the term used to describe the several ways in which plants can grown without soil, by use of an inert medium where plant can take essential nutrients, either from water to which is added a nutrient solution or from organic materials that exists in the medium. These media can be gravel, sand, peat, vermiculate, prelite, sawdust or other plant matters (organic components). Therefore, the process in which aquatic weeds are dumped on water to construct floating bed or artificial island for producing agricultural crops is called hydroponics or Floating Garden. This farming system is locally also known as *Baira* or *Geto* or *Bed* or *Dhap*. Different types of vegetables, seedlings and flowers are grown on it (Haq *et al.*, 2002).

Global climatic change issue is a concern worldwide. The impact of climatic change on flood plain poor country like Bangladesh have also been highlighted repeatedly as cautionary signals to equip people to face. The eventual consequence when it comes. Many people think that 35% of the land in Bangladesh would go under water in the near future due to sea level rise caused by global warming and green house effect. It is worth mentioning here that in south west region the water logging is increasing day by day as a negative impact of Costal Embankment Project (CEP) At present the south west region of Bangladesh exists about 4,24,538 acres of natural and manmade wetlands. Out of these about 45,512 acres of wetland might be fully used for the practice of soil-less agriculture. Nearly 5,000 acres of wetland are using for this purpose (Haq *et al.*, 2002). The soil-less agriculture be able to turned "the curse" into well-to-do site by making these vast wetlands productive and ensure food security that direct livelihood sustainability of the wetlands peoples (World Bank, 2000).

Materials and Methods

Materials

Floating bed or *Dhap* is constructed based on the availability of different aquatic weeds and other plant materials at different locations. Mainly Water hyacinths (*Euchornia crassipes*), Aman paddy stub, Reeds, different decomposing aquatic weeds *i.e.* Water letuce (*Pistia stratiotes*), Duckweed, *Najas graminea*, *Salvinia spp*, *Potamogeton alpinus*, Coconut husk, Bamboo, Country boat, Chopper, etc. are used in constructing floating beds.

Cultivation Procedure

Water hyacinth profusely grown in wetlands such as low-lying areas closed rivers, canal, lagoons, etc, floating bed can be prepared in any depth of water and carried this bed by rowing to the farmer's desired site. 5 decimal areas of water hyacinths is needed to construct 1 decimal floating bed

To begin with, farmers put a long bamboo as desired by him, on the mass of the fully matured water hyacinths (immature water hyacinths decompose faster) then a single man stand on the bamboo over the mass of water hyacinths and balances himself to pull the water hyacinths from the both sides of the bamboo and flattened them under foot. In this process, he proceeds towards the end of the bamboo. This process is continued until the desired height and length of the bed is attained. When the construction of floating bed is completed, the bamboo is removed from the bed. Farmers again dump water hyacinths after 7-10 days later from the first dumping and then the bed remains for decomposition before the commencement of plantation. Sometimes farmers use semi-decomposed aquatic plants such as water lettuce, Duck weed, *Najas* spp, *Salvinia* spp and immature water hyacinths etc. on the top of the bed for (i) decomposing the top of the bed quickly (ii) making available nutrient for seedlings (iii) avoiding extra evaporation from bed and iv) making an ideal ground for setting germinated seeds and different crops (Haq *et al.*, 2002).

As the first layer of water hyacinths do not decompose quickly, the floating bed can easily keeps its buoyancy. First layer of water hyacinths acts as the base of floating bed and it maintains the stability, buoyancy and thickness of the bed. Generally, the top of the floating bed needs 15-20 days to be totally decomposed for sowing seed or planting seedlings. It is remarkable that the newly constructed floating bed can be cultivated from the first day by spreading compost (natural decomposing material) manure thickly on the bed. There after seed or *Tema* (A ball made of compost manure and aquatic creepers in which seeds are inserted for safety germination) are placed on the bed. The *Tema* is locally called as *Dalua*, *Ball*, etc.

Cropping Pattern

Vegetables and seedling are the main crops of this farming system. In this system about 23 types of vegetables and 5 types of spices are grown. The vegetables and seedlings which are raised on floating bed are Ladies finger (Okra), Cucumber, Ridged gourd, Bitter gourd, Snake gourd, Amaranth, Brinjal (egg plant), Pumpkin, Indian spinach, Taro, Wax gourd, Turmeric, etc. as monsoon crops. Apart from these Spinach, Bottle gourd, Yard long bean. Bean, Tomato, Potato, Cauliflower, Cabbage, Kohlrabi, Turnip, Radish, Carrot, Ginger, Onion, Chili, Garlic, etc. are belonging to winter crops. Some vegetables are grown on the bed all the year round rotationally.

Cost benefit analysis of floating garden (hydroponics)

Cost for 10 floating beds (each bed is 45 ft × 6 ft × 3 ft) only in floating condition

Sl.No.	Cost head	Quantity	Unit cost (Tk)	Total
1	Construction of floating beds	60 man days labor	50.00	3000.00
2	Collection of raw material (weeds)	20 man days labor	50.00	1000.00
3	Seed and or seedling purchase		60.00	600.00
4	Bamboo, rope, crop harvesting and maintenance		100.00	1000.00
	Total cost			**5,600.00**

Income from 10 floating beds (each bed is 45 ft × 6 ft × 3 ft) only in monsoon period

Sl.No.	Income head	Quantity (kg)	Unit income (Tk)	Total
1	Ladies finger (okra)	1,800	5.00/kg	9000.00
2	Ridged gourd	400	6.00/kg	2400.00
3	Amaranth (red colored)	600	5.00/kg	3000.00
4	Others (Taro, Indian spinach etc.)	150	4.00/kg	600.00
5	Organic compost manure	30,000	0.20/kg	6000.00
	Total Income			**21,000.00**

Total Benefit = 21,000.00 Tk – 5,600.00 Tk = 15,400.00 Tk (BDT). If the farmers do not sell the organic manure (decomposed water hyacinths and aquatic weeds) and keep it for the next year application for better yield, he will gain initially only = 21,000.00 Tk – (5,600.00 + 6,000.00) Tk = 9,400.00 Tk. If the farmers contribute his/her own labor in this farming system, he/she will save 4,000.00 Tk and net benefit will be 9,400.00 + 4,000.00 = 13,400.00 Tk excluding the price of compost manure.

Apart from this when the water recedes the floating bed (organic manure) tends to be flatten on soil. The beds then transformed into semi-solid minute particles and spread over the soil. Crops belonging to winter can be raised on this soil without any tillage and fertilizer. In this way, more benefit can be achieved with less input from non-floating stage (Haq *et al.*, LEISA 2004).

Discussion

Comparative Advantage of Hydroponics as a Means of Wetland Resource Management

Soil-less agriculture for appropriate utilization of wetlands resources

Bangladesh constitutes about 50% of its plain land. Besides, a vast area of the country remains submerged under water for a long period during the monsoon.(Khan *et al.*,1994.) So far, these wetlands, including waterlogged areas were looked upon to be the crisis regions of the country, as no terrestrial crops could grow there but mostly covers with aquatic invasive plants. To increase agricultural production, large numbers of drainages and irrigation project was introduce throughout Bangladesh, including some relief activities was undertaken by both Govt. and NGOs since the 60s. For a short time, such efforts produce some benefits; ultimately alter the normal functioning of the nature. As a result, water logging and flooding have been alarmingly increasing day by day. If Soil less agriculture can disseminate to different part of Bangladesh then policy maker would change their attitude. In these connections, we need to change our terrestrial agriculture base approach. We have to consider about the resources of wetlands and need national management plane of it based on Soil-less agriculture (Ghosal and Haq, 2000).

Restoration of wetlands ecosystem through management of aquatic invasive plants

Aquatic invasive species are considered to be the second largest reason for biodiversity loss worldwide www.bcpc.org/invasive. The introduction of Water Hyacinth to Africa's wetlands alone has caused billions of dollars of damage. (New Agriculturist, 2004) Florida (USA) likely has the largest aquatic plant management program in the World, spending more than $70 million annually (Tyler, 2004). In Bangladesh, large magnitude of water hyacinths and other aquatic invasive plant profusely grown and infested water ways, closed river, oxbow-lake, seasonal and perennial water bodies etc that incur huge damage in fisheries, agriculture create pollution and hinder navigation. Large-scale introduction of Soil-less agriculture can capture those problems with win- win basis.

Environment friendly intensive farming system

The increase in yields per acre under hydroponics cultivation is striking compared to the terrestrials' agriculture. Hydroponics, on the contrary, being labor-intensive one, offers abundant employment opportunity in the populous county like Bangladesh.

Hydroponics increases resource base of the community

Hydroponics could have a positive impact on open water fisheries and other non-traditional off-season agricultural crops (Agro-biodiversity) strengthening the resource base of the community. Hydroponics farming is a sustainable form of agricultural/economic development in specific Bangladesh context, which contribute positively to the conservation of biological diversity.

Hydroponics provides shelter during flood

During flood people can use floating mats (*Dhap*), to some extent, to save their dwellings as well as their household articles including domestic animals from the damaging effect of waves in the vast water-body (mainly inundated area by flood) in those areas.

Hydroponics enhances soil quality

Approximately 6 ton/decimal amount of compost material produced through soil-less agriculture that can be used in agricultural land as compost manure for enriching organic content of the soil. The compost materials have a huge Market potentiality, as soil degradation in Bangladesh due to loss of organic matter is significant

Conflict resolution with fisheries

Wetland is the natural breeding ground of a large number of fish during rainy season. Besides, under 3rd and 4th Fisheries Program Fisheries Department of Govt. of Bangladesh release huge number of fingerlings of different carp species during monsoon for livelihood development of community. Fingerlings take 3 to 4 months for its growth and maturity. For the shake of the growth of these fingerlings, it is necessary to discontinue capture fishery during this period. On the other hand, during monsoon there are immense crisis for employment, most of the wetland people depends on fisheries. This creates tension and conflict with law and order enforcing authority in the above situation.

Globalization and resources conserving agriculture.

The impact of free trade shows that most small farmers cannot compete on the global market;. The emerging alternative to "globalization" is increasingly called "localization". As an alternative development approach,

it gives priority to endogenous development – building on local concepts and resources, but does not exclude exogenous development – building on solution from outside. The local economies make better use of available natural and social resources by minimizing the use of external resources. Endogenous development of local economies is not backward at all. It is an artful and knowledge intensive way of living, as demonstrated, in principle, by many traditional "Forgotten agricultures," hydroponics would help farmers to design strategies to navigate their lives in a fast globalizing world with bear minimum external input in this farming system and conserving their natural resources (Haq *et al.*, 2002).

Ensures women participation

Women have specific contribution in this farming system by preparing *Tema*. They also participate in other activities of cultivation like maintenance, harvesting etc. along with their household work.

Water and nutrients are conserved

A properly designed hydroponics system uses much less water and nutrients than conventional soil based culture. This is because the nutrients can be recycled through the system. This advantage is especially significant as it can lead to a reduction in pollution of the land and the stream with high level of run off nutrients (Haq *et al.*, 2002 and Tse, 1995).

Pest and disease problems are reduced

The chance of soil borne disease is greatly reduced with hydroponics, as it is a soil less culture (Tse, 1995).

Transplanting shock is reduced for seedlings

In hydroponics cultivation, seedlings can be easily raised in *Tema* (a ball of compost manure for safety germination of seeds). This *Tema* then be transplanted directly in to the hydroponics beds without the need to prick out the plants as in the case of soil media. Hydroponics, therefore, shortens the propagation time and reduces the transplant shock in young seedlings (Tse, 1995).

In hydroponics culture, plant nutrients and water are readily available in sufficient quantities year round, this allows higher density planting, and it is even possible to grow plants in multilevel with hydroponics.

Recycling of waste water resource

The water hyacinth is proved very effective filtration system for cleansing wastewater containing a complex chemical mixture. Water hyacinth poses an extensive root system, which allows them to feed directly from the aqueous medium, extracting chemicals and nutrients rapidly and effectively.

Another feature is the plants tremendously high growth rate, capable of producing 17.5 metric tons of wet biomass per hectare per day under ideal growing condition. Water hyacinths accumulated heavy metals to concentrations several hundred times the initial levels. Water hyacinths biological filtration capability is very high and reduces BOD up to 90% (Mitsch, 1989 and Wolverton, 1976). The chemical oxygen demand (COD) has also been reduced by 83-92%. In Bangladesh, large numbers of water bodies exist in and around the mega cities, urban and peri urban areas which receive large amount of waste water that create environmental hazards. These wastewater can be treated as usable water using water hyacinth in soil less agriculture (Saha *et al.*, 2000).

Conclusion

Extreme events such as heavy rains and droughts are the most destructive forms of weather, and the frequency and duration of these events are likely to increase as the climate continues to change. Droughts and floods occur naturally around the world, for example in association with El Niño events, but are likely to become more severe, causing water management to become an even more critical problem in the future.

Bangladesh has one of the most densely populated, low lying coastal zones in the world, with 30 million people living within 1 meter elevation from the high tide level. It will face two prong situations such as northern dry part will face further increasingly high temperature and low rainfall leading to drought condition due to climate change. People in those areas are building monsoon flood water retention initiatives which are potential sites for introducing soil less floating agriculture *known as hydroponics*. On the other hand, southern part toward the coastal areas will experience more flood, water stagnation for a longer duration which can be brought under productive and adaptive farming system through practicing *hydroponics*. Rising temperature, one of the effects of climate change that impede land based agricultural production (Brown, 2004) is not a factor for soil less floating agriculture.

It may be possible for global agricultural production to keep pace with increasing demand over the next 50-100 years if adequate adaptations are made, but there are likely to be difficulties in country like Bangladesh. Development of effective adaptation strategies requires community participation in response strategy development and recognition of multiple stresses on sustainable management of resources. Since early signs of climate change already have been observed and may become more prominent over time, it is time to take measure to design and implement adaptations like *hydroponics*. Otherwise it may be too late to avoid disasters.

Hydroponics are found to be the best practice as well as win-win kind of farming system which provided high agricultural yield without application of chemical fertilizer and pesticide and the process itself also facilitate to improve the degraded wetlands and lands. This farming system ensures scope of participation for both male and female, farmer and scientist to overcome the emerging challenges due to climate change at the grassroots level.

Since the climate change impact is eco-specific, it is pertinent to develop and demonstrate location specific available resource based hydroponics models. Before recommending widespread dissemination, it is critical need to document the procedural guideline based on standardizing the indigenous technique of hydroponics through different Research & Development initiative including use of Participatory Technology Development tools.

References

1. A. K. Saha, M.S. Shah and Y.A. Jamadar (2000). Nutrient Recovery and Treatment of Fish Processing Plant Effluent by Water Hyacinth. In Waste Recycling and Resource Management in the Developing World: Ecological Engineering Approach (501-504). Edited by B. B. Jana, R. D. Banerjee, B. Guterstam and J. Heeb. Jointly published by University of Kalyani, India and International Ecological Engineering Society, Switzerland.

2. Brown, Lester R. (2004). Outgrowing the Earth, W.W. Norton & Company, New York(pp. 118-132) .

3. Haq, A. H. M. Rezaul , Asaduzzaman, M. and T. K. Ghosal (2002). Soil-less Agriculture in Bangladesh. 111 pp., A Grameen Trust, Bangladesh Publication under the component of Research for Poverty Alleviation. Grameen Bank Bhaban, Mirpur 2, Dhaka 1216, g-trust@grameen.com

4. Haq, A.H.M. Rezaul, T.K.Ghosal and Pritam Ghosh (2004). Constraints for the dissemination of Hydroponics in the South-West region of Bangladesh. WRDS publication, February 2004, p-50, Khulna, Bangladesh. wrds@bttb.net.bd, wetlandbd@gmail.com

5. Haq, A.H.M. Rezaul, T.K.Ghosal and Pritam Ghosh (2004). Cultivating Wetland in Bangladesh, LEISA published by ILEIA, The Netherlands, Vol: 20 No. 04 pp: 18-20. http:// www.leisa.info

6. Khan M Salar, Enamul Haq, Saleemul Haq, A Atiq Rahman, S M A Rashid and Helal Ahmed (1994). Wetlands of Bangladesh. Published by Bangladesh Center for Advanced Studies (BCAS).

7. Leow Atomic Chuan Tse (1995). A guide to Hydroponics, Edited by Joyce T S Foo, published by Singapore Science Center.

8. Mitsch, W. J., (1977). Water hyacinth (*Eichhornia crassipes*) Nutrient Uptake and Metabolism in a North Central Florida Marsh. *Arch. Hydrobiol.*, 81:188-210.

9. New Agriculturist (2004). Bogged down with aquatic weeds, CD-Resources. Bi-monthly on line journal. Issue: January 2000 to April 2004. WREN media. www.new-agri.co.uk.

10. T. K. Ghosal and A.H.M.R. Haq (2000). Hydroponics-an approach for wetland management. In Waste Recycling and Resource Management in the Developing World: Ecological Engineering Approach (511-513). Edited by B. B. Jana, R. D. Banerjee, B. Guterstam and J. Heeb. Jointly published by University of Kalyani, India and International Ecological Engineering Society, Switzerland.

11. Tyler. J. Koschnick (2004). Aquaphyte weed training course. *AQUAPHYTE* vol-24 no.-1, summer 2004.

12. Wolverton, B.C. and R.C. McDonald (1976). Water Hyacinths for upgrading sewage lagoons to meet advanced wastewater treatment standards: Part II. NASA Technical Memorandum TM-X-72730.

13. World Bank (2000). Bangladesh, Climate Change and Sustainable Development, Report no. 1104 BD, World Bank, Dhaka.

14. "Alternative Livelihood Approach in Ecofragile Zones of Odisha (Seminar Report) organized by BIRD (NABARD),Lucknow in Puri"(2011).

Sustainable Development of Natural Resources and   *Pages* **88-93**
Wildlife Conservation
Editor: **Ashwani Kumar Dubey**
Published by: **REGENCY PUBLICATIONS, NEW DELHI**

13
Mathematical Model for Phytoplankton Growth

Usha Pancholi*, Prahlad Dube# and R. K. Sharma**

ABSTRACT

Phytoplanktons are producers in aquatic ecosystem and are sources of energy to higher levels (consumers) of food chains and food webs. A simple mathematical model is described in the present paper called nutrient – phytoplankton model. This model tries to understand the dynamics of nutrient driven phytoplankton growth. In the literature surveyed so far, a large number of complicated simulation studies are reported. It is found in actual (realistic) field studies that the growth of the Phytoplanktons is strongly controlled by the nutrients rather than by consumer of higher trophic level (e.g. population densities). The present model is based on the assumption that the growth is accelerated when nutrients exceed a certain level (ecologically saying, that is, oligotrophic water resource turn into eutrophic one). This threshold effect should be generic to simulation models.

Key words: *Nutrients, nutrient – phytoplankton model, phytoplanktons, lentic water bodies.*

Introduction

Phytoplanktons are photosynthetic, tiny plants floating in the surface water in the lentic (non-flowing, stagnant) water bodies, rivers (lotic resources) and oceans. Phytoplankton community occupies the lowermost and starting part of almost all food chains (detritus food chains are exception)

* Department of Mathematics, Government College, Kota-324009, (Rajasthan).
# Department of Zoology, Government College, Kota-324009, (Rajasthan).
**Department of Mathematics, Government College, Bundi, (Rajasthan).

and thus is very important functionally. Most of the growth studies of planktons are performed in laboratories and using variety of methods and culture techniques. Therefore, formulation of a mathematical model to understand and explain the phytoplankton growth is difficult. Secondly, most of the experiments are performed in steady-state conditions. A mathematical model, in such situation, can provide the means for keeping track of the complexities and for generalizing the results obtained to transient, non-steady-state conditions. Natural ecosystems contrary to the laboratory culture tanks are characterized by non-steady conditions.

Mathematical models of various aspects of phytoplankton dynamics were formulated earlier. In the literature, the earliest attempt to develop a model of phytoplankton productivity was the work of Fleming (1939) which was based on Volterra equations. Sverdrup *et al.* (1942) and Riley (1946) included light intensity effects on production as well as factors such as temperature – dependent respiration and nutrition depletion. The physical processes such as sinking rate, diffusivity and vertical instability were included in later models (Steele 1958, 1962 and Patten 1963, 1965, 1966). Patten (1968) reviewed early models extensively.

The most commonly used relationship between phytoplankton growth concentrations originates with works of Caperon (1967) and of Dugdale (1967), which used Michaelis-Menten kinetics to describe nitrogen uptake rate by algae as a function of ambient nutrient concentration. Droop (1973) has argued that growth actually responds to size of internal nitrogen pool rather than directly to the external nutrient concentration. Steel and Henderson (1992) have developed model formulations in a simple nitrogen-phytoplankton-zooplankton (NPZ) system that focused on primarily grazing. Fashman *et al.* (1990) published a model (FDM) which is a well documented and well thought-out model. In this, authors compared its results with field observations. This approach made the basis for present paper.

The Growth Process and Its Equation

The phytoplanktons grow by absorbing and utilizing the nutrients (mainly nitrogen in the form of nitrite, nitrate and ammonium and phosphate) through a complex metabolism in their cells and tissues. This can be expressed in the form of following equation:

Uptake   Metabolism

Nutrients $\longrightarrow$ Absorption and utilization $\longrightarrow$ Growth

(Nitrite, nitrate, ammonium and phosphate)   (Increase in Number/biomass)

The Model

The two nutrient parts for nitrogen and phosphorous represent the fractions that a cell uses for growth. The phytoplankton part represents the numbers. Each of the parts is expressed in gm per cubic meter and numbers per ml of water respectively. The model can be represented by —

$$dP/dt = \beta_0\,\alpha_1\,PN - \beta_1\,P - \beta_2\,P^2 - \beta_3\,PZ \tag{1}$$

N   : Nutrients

P   : Phytoplanktons

$\beta_0\,\alpha_1\,PN$: Production of Phytoplanktons limitation of N

$\beta_1 P$: Natural depletion rate of Phytoplanktons

$\beta_2\,P^2$: Death of Phytoplanktons

$\beta_3\,PZ$: Grazing of Phytoplanktons by zooplanktons

Phytoplankton production is expressed by –

$$prPC = \mu f(I)\ f(T)\ f(N,P)\ FCrd, \tag{2}$$

where $\imath$ – maximum growth rate of phytoplankton, $f(I)$ – dependence on light availability, $f(T)$ – temperature dependence, $f(N,P)$ – nutrient function, FC – correction factor for dark reaction, rd – relative day length. The nutrient function is calculated as

$$f(N,P) = \frac{2}{\dfrac{1}{NF} + \dfrac{1}{PF}} \tag{3}$$

where f(N,P) = nutrient function, NF = nitrogen function, and PF = phosphorus function.

The effect of nutrients on production/growth is demonstrated by Figures 13.1-13.3 and tables 13.1-13.3. It is evident that there is an increase in phytoplanktons and during the blooms it appears to limit the growth severely. The model for growth was tested on this pond and the results are shown in figure 13.3.

Table 13.1: Theoretical No. of phytoplankton in a perennial pond in Kota (Rajasthan) India

Sl.No.	Time (Days)	No. of phytoplanktons (thousands per ml)
01	001	000
02	071	013
03	141	036
04	221	150
05	281	050
06	323	025
07	365	010

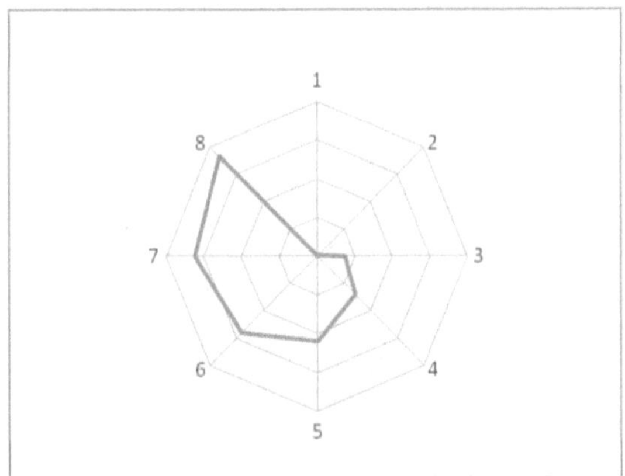

Fig. 13.1: Diagrammatic representation of theoretical No. of phytoplankton in a perennial pond in Kota (Rajasthan) India.

Table 13.2: Total count of phytoplankton (Observed data) in a perennial pond in Kota (Rajasthan) India.

Month	No. of Phytoplanktons
01	853
02	1047
03	2471
04	1633
05	2460
06	1829
07	1005
08	2812
09	2646
10	1893

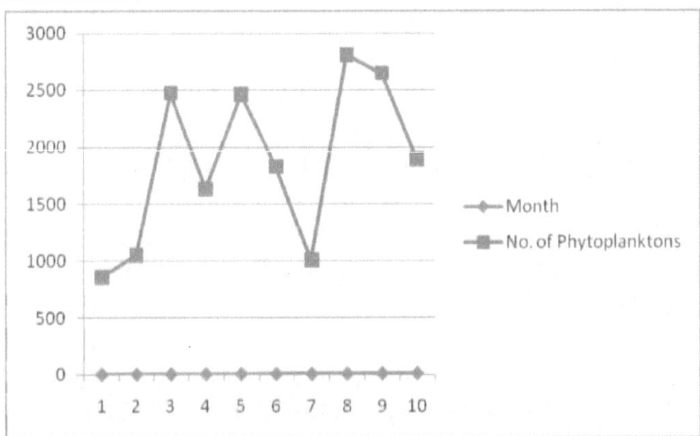

Fig. 13.1: Diagrammatic representation of Total count of phytoplankton (Observed data) of phytoplankton in a perennial pond in Kota (Rajasthan) India.

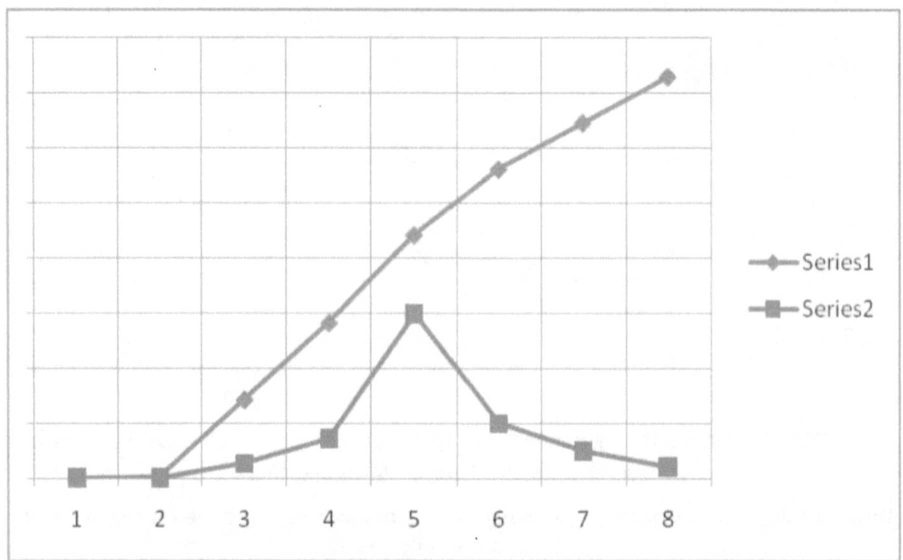

Fig. 13.3: Diagrammatic representation of comparison of theoretical No. and Total count of phytoplankton (Observed data) a perennial pond in Kota (Rajasthan) India.

Conclusion

The predicted phytoplankton numbers are plotted against observed numbers during the time period and the agreement is sufficiently good. Recent advances in phytoplankton growth studies permit realistic mathematical modeling of phytoplankton growth. Nutrient uptake can be represented as separate process. The excretion (of nitrogen) and non-predatory mortality can be modeled as functions of other factors (such as temperature, light and limiting factors).

References

Caperon, J. (1967). Population growth in micro-organisms limited by food supply. *Ecology*, 48:715-722.

Droop, M. R., (1973). Some thoughts on nutrient limitation in algae. *J. Phycol.*, 9:264-272.

Dugdale, R. C., (1967). Nutrient limitations in sea: Dynamics, identification and significance. Limnolo. *Oceanogr.*, 12:685-695.

Fashman, M. J. R., Ducklow, H. W. and McKelvie, S. M. (1990). A Nitrogen based model of plankton dynamics in the oceanic mixed layer. *J. Mar. Res.*, 34:591-639.

Fleming, R. H. (1939). The controle of diatom population by grazing. Rapport des Reunions Conseil International pour Explorations de la Mer 14:210-217.

Patten, B. C. (1963). The information concept in Ecology; some aspects of information-gathering behavior in plankton. Pages 140-172 *in* W. S. Fields and W. Abbott, editors. Information storage and neural control. Charls C. Thomas, Spring field, Illinois, USA.

Patten, B. C. (1965). Community organization and energy relationships in plankton. Oak Ridge National Laboratory Report. 3634:1-60.

Patten, B. C. (1966). The biocoenogenetic process in an estuarine phytoplankton community. Oak Ridge National Laboratory Report. 3946:1-97.

Patten, B.C. (1968). Mathematical models of plankton production. *International Revue der Gesamten Hydrobiologie* 53:357-408.

Riley, G.A. (1946). Factors controlling phytoplankton populations on Georges bank, *Journal of Marine Research*, 6:54-73.

Steele, J.H. (1958). Plant production in the northern North Sea. Rapports des Reunions Conseil International pour Explorations de la Mer 144:79-84.

Steele, J.H. (1962). Environmental control of photosynthesis in the Sea. *Limnology and Oceanography* 7:137-150.

Steel, J. H. and Henderson, E. W. (1992). The role of predation in plankton models. *J. Plankton Res.* 14:157-172.

Sverdrup, H.U.M.W. Johanson and R.H. Fleming. (1942) . The Oceans. Prentice Hall, Englewood Cliffs, New Jersey, USA.

Sustainable Development of Natural Resources and   *Pages* **94-102**
Wildlife Conservation
Editor: **Ashwani Kumar Dubey**
Published by: **REGENCY PUBLICATIONS, NEW DELHI**

14

Bird Diversity at Chhatarpur District, Madhya Pradesh, India

Ashwani Kumar Dubey

ABSTRACT

The incredible numbers of bird species demonstrate amazing evolutionary adaptations, and by learning how birds are able to adapt throughout the world. We can begin to adapt our own behaviors to live in our world, rather than to force our world into an artificial and unsustainable mold. Biodiversity is not evenly distributed across the earth. Many reasons the bird population is declining is due to deforestation, population explosion, indiscriminate use of pesticides, hunting, and destruction of habitat, pollution and contaminated water. Therefore, in the present investigation preliminary observation of Birds carried out in Chhatarpur district Madhya Pradesh, India.

Introduction

Birds are an integral part of the ecosystem and have importance for eco-balance. Migratory birds can gain a better understanding of seasonal climate changes. By conserving birds and protecting their habitats we can continue to gain insights from our birds friends. Birds are one of the most

Research & Development Unit, Godavari Academy of Science and Technology, Environment & Social Welfare Society Chhatarpur 471001 India
Corresponding author: Email: ashwani_0326@yahoo.com; Phone: +91-9425143654

populous life forms on the planet, and that biodiversity leads to a richness of life and beauty. It may be influenced by biography (Karr 1976). Some landscape exhibit high richness in biological diversity where others show an impoverished flora and fauna. Various scientist have been conducted to look at bird diversity in South Indian Forest (Joshua and Johnsing 1986, Pramod et al 1997, Kunte *et al* 1999), relationship between birds species diversity and vegetation (Able 1976, Terbrgh 1985, Hawkins 1999, Joshi et al 2012), factors responsible for species distribution (Lee 2004, Bhatt and Joshi, 2011) bird diversity (Dodia and Dhadhal, 2010).

The numerous observations by amateur and professional bird watchers may support the idea of the value of habitat of bird diversity conservation. The world famous monuments Khajuraho is located in the district Chhatarpur and as a result high flow of foreign tourist. Due to the tourism industries, many environmental issues are rising up in the Chhatarpur district. Nevertheless there is lack information of scientific documentation and preventive measure on bird diversity conservation in this area.

Many reasons the bird population is declining is due to deforestation, population explosion, indiscriminate use of pesticides, hunting, and destruction of habitat, pollution and contaminated water. Therefore, in the present investigation preliminary observation of Birds carried out in Chhatarpur district Madhya Pradesh, India.

Materials and Methods

Human beings affect the survival of birds by modifying their habitats. Aim of the present study providing a comprehensive list of the bird's species of Chhatarpur District Madhya Pradesh.

Study Area

Chhatarpur geographically located with longitudes and latitudes of 24^006 and 25^020 on North 78^059 and 80^026 on East respectively with approximate 182 meter above means sea level experiencing a annual rainfall of 1000-1200 mm. Average climatic temperature in winter season (October to January) 10-27^{0}C, summer season (February to June) 29-48^{0}C and rainy season (June to September) 19-30^{0}C. The total area of the Chhatarpur district is about 8,687 km^2. Chhatarpur district is bounded by Uttar Pradesh state of the North and the Madhya Pradesh district of Panna to the East, Damoh to the South Sagar to the Southwest, and Tikamgarh to the West. The district is divided into eleven tahseel *viz.* Badamalahra, Bakswaha, Chandala, Chhatarpur, Gaurihar, Ghuwara, Lovekushnagar (Laundi), Maharajpur, Nowgong, Rajnagar and Vijawar.

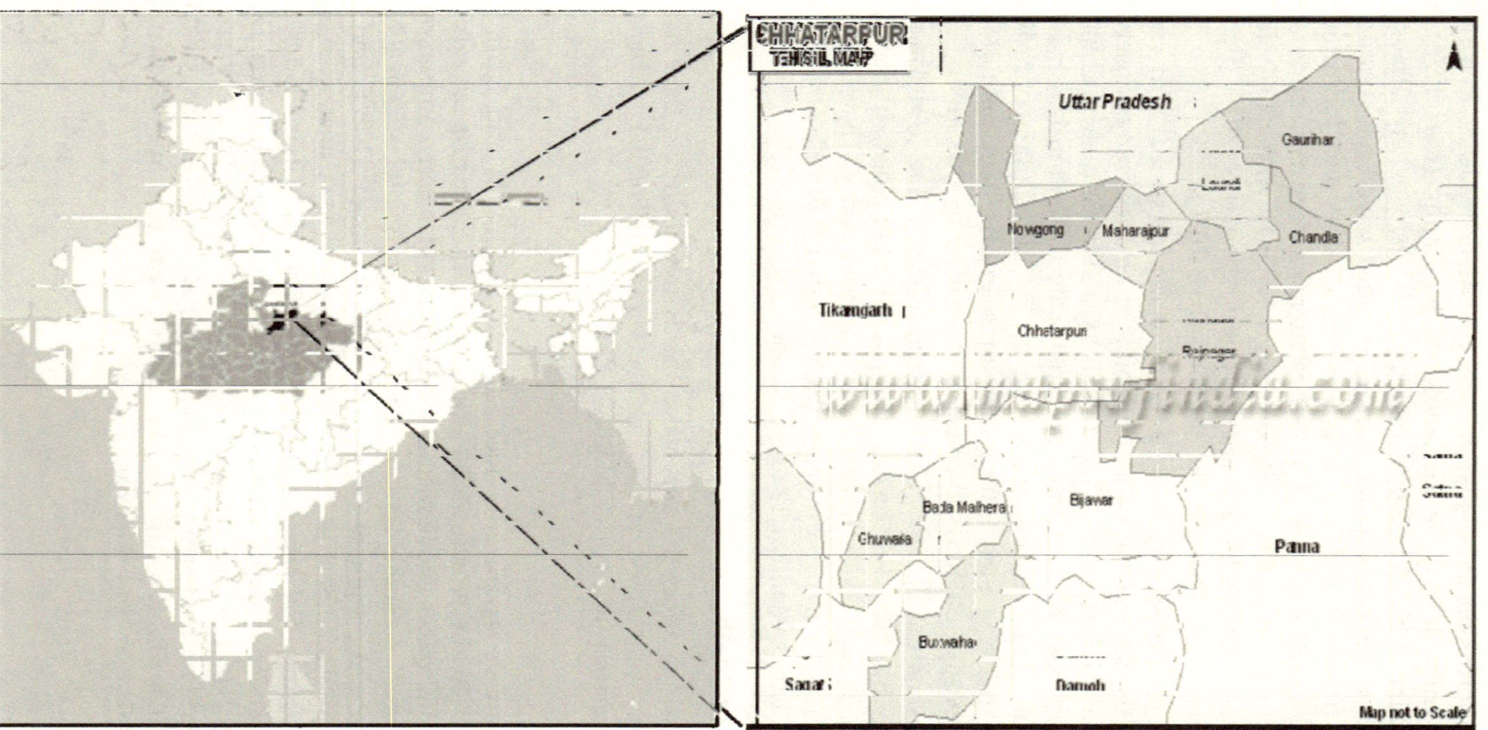

Fig. 14.1: Map showing geographic location of Chhatarpur District Madhya Pradesh, India.

Animal Observations

In the present study we considered the Chhatarpur district Madhya Pradesh for bird diversity because world famous monuments Khajuraho is located in this district and as a result high flow of foreign tourist. Almost worldwide bird lovers are visited this place. Field study was conducted for three year from June 2010 to August 2013 with the aim of providing comprehensive list of the bird's species. The birds have been monitored using Variable Circular Plot method (de Filippo *et al.*, 1996). Field identifications were carried out with the help of field guide (Ali and Ripely, 1983) Field binoculars (7×50) were used to observe the birds. Another aspect keep in consideration, the activity of birds during sunrise and sunset, and monitoring of transects done. The scientific names of birds are according to after (Manakadan and Pittie, 2001).

Results and Discussion

Diverse populations of birds have been identified in the Chhatarpur district at different selected geographic site either as breeding population, winter, rainy and summer visitor or migration. Aquatic birds observed in randomly selected ponds; Prem sagar Khajuraho, Jalsena Rajnagar, Godavari Fisheries Demonstration Centre Nahdora, Sankat mochan Chhatarpur, Dam; Banisagar, Budha, Gangaun, Ranguan, Bariyarpur and in the bank of River Ken and Dhasan. Total 73 birds species belonging to twenty eight families were reported in and around Chhatarpur district (Table 14.1).

Bird diversity is critical study. We share more than 10,000 species of birds. Study of bird diversity means understanding ecology. Birds are an integral part of the ecosystem and serve many important purposes. Biodiversity, we can better understand the relationships between all living organisms and how the interactions of those relationships can affect humans directly. Species richness decreased with increasing urbanization (Melles *et al.*, 2003). Urbam birds communities are usually characterized by the dominance of a few species (Beissinger and Osborne, 1982, Marzluff, 2001) and most of the species making up the communities are introduced. Most of the researchers studied on the base of urban avoiders, suburban adaptive and urban exploiters (Blair 1996, McKinney, 2002). Localized extensions may be a result of species invasions by non-native species (Kowarik 1995, Blair 2004). A low spatial variation of urban bird communities is expected and may probably result to more similar bird communities all over the world (Jokimaki et al 1996).

Bird can also highlight the diversity of different habitats. All birds cannot live in the same habitats, and understanding the needs and requirements of different species can lead us to have more compassionate tolerance for different environment condition. It is also indicators of pollutants and habitat variety. In Chhatarpur district Motacilla alba can be

Table 14.1: List of the Birds species recorded during the study (2011-2013)

Sl.No.	Family	Scientific Name	Common Name	Conservation Status (IUCN)	Conservation Status (IWPA)
01	Acciptridae	*Milvus migrans*	Black kite		Schedule IV
02		*Gyps bengalensis*	White rumped vulture	CR	Schedule IV
03		*Gypus fulvus*	Vulture		
04		*Gypus indiacus*	Indian vulture		
05		*Neophron percnopterus*	Egyptian vulture		
06		*Hieraaetus pennatus**	eagle		Schedule IV
07		*Accipiter virgatus**	hawk		Schedule IV
08		*Aviceda jerdoni*			Schedule IV
09	Alaudidae	*Calandrealls raytl*			Schedule IV
10	Alcedinidae	*Halcyon smyrnensis*			Schedule IV
11		*Alcedo atthis*	Common kingfisher		Schedule IV
12		*Halcyon pileata*			Schedule IV
13	Anatidae	*Anas acuta*			Schedule IV
14		*Anas peocilorhyncha*	Indian spot billed duck		Schedule IV
15		*Aythya fernia*	Common pochard		Schedule IV
16		*Mergus merganser*			Schedule IV
17	Apodidae	*Bubulcus ibis*	Cattle egret		Schedule IV
18		*Apus affinis*	House swift		Schedule IV
19	Bucerotidae	*Anthracoceros malabaricus*			Schedule IV
20	Campephagidae	*Pericrocotus ethologus*	Long tailed minivet		Schedule IV
21	Ciconidae	*Ciconia episcopus*	Wooly-necked-stork		Schedule IV
22	Charadriidae	*Vanellus indicus*	Red-wattled lapwing		Schedule
23		*Ciconia nigra*	Black stork		
24	Columbiadae	*Columba livia*	Blue rock pigon		Schedule IV

Contd...

Table 14.1: *Contd...*

Sl.No.	Family	Scientific Name	Common Name	Conservation Status (IUCN)	Conservation Status (IWPA)
25	Coraciidae	*Coracias benghalensis*	Indian roller		Schedule IV
26	Corvidae	*Corvus splendens*	Common crow		Schedule IV
27		*Corvus macrorhynchos*	Large billed crow		Schedule IV
28	Cuculiade	*Eudynamys scolopacea*	Asian koel		Schedule IV
29		*Centropus sinensis*	Greater coucal		Schedule IV
30		*Cuculus canorus*	Common cockoo		Schedule IV
31		*Centropus micropterus*	Indian cuckoo		Schedule IV
32	Egrets	*Ardea goliah*			Schedule IV
33		*Ardea cinerae*			Schedule IV
34		*Ardea alba*			Schedule IV
35		*Ardeola grayii*			Schedule IV
36		*Bulbulcus ibis*			Schedule IV
37		*Egretta ganzella*	Little egret		Schedule IV
38		*Egretta intermedia*	Intermediate egrate		
39		*Egretta aalba*	Great egret		
40	Gruidae	*Grus grus*			Schedule IV
41		*Bubo coromandus*	Western cattle egret		
42		*Grus antigone*			Schedule IV
43	Hirundinidae	*Hirundo rupestris*			Schedule IV
44	Motacillidae	*Motacilla alba*	Khanjan		Schedule IV
45	Muscicapidae	*Terpsiphone paradisi*	Asian paradise fly catcher		Schedule IV
46	Passeridae	*Passer domesticus*	House sparrow		Schedule IV
47		*Ploceus philippinus*	Baya		Schedule IV
48	Phasianidae	*Pavo cristatus*	Indian peafowl		Schedule IV
49		*Francolinus pictus*	Kala Teetar		Schedule IV

Contd...

Table 14.1: *Contd...*

Sl.No.	Family	Scientific Name	Common Name	Conservation Status (IUCN)	Conservation Status (IWPA)
50		*Francolinus pondicerianus*	Ram Teetar		Schedule IV
51		*Gallus gallus*	Red jungle fowl		Schedule IV
52		*Gallo perdix spadicea*	Grey jungle fowl		
53		*Gallus sonneratti*			Schedule IV
54	Picidae	*Dendrocopos mahrattensis*	Yellow crowned woodpecker		Schedule IV
55		*Brachypternus bengalensis*	Woodpecker		Schedule IV
56		*Chrysocolaptes festivus*	Woodpecker		Schedule IV
57	Prinia warblers	*Prinia sociaiis*			Schedule IV
58		*Chrysocolaptes festivus*			Schedule IV
59	Psittacidae	*Psittacula eupatria*	Alexandrine parakeets		Schedule IV
60		*Psittacula krameri*	Rose ring parakeets		Schedule IV
61		*Psittacula cyanocephala*	Plum headed parakeet		
62	Pycnonotidae	*Pycnonotus cafer*	Red vented bulbul		Schedule IV
63		*Pycnonotus melanicterus*	Bulbul		Schedule IV
64	Sturnidae	*Acridotheres ginginianus*	Bank myna		Schedule IV
65		*Acridotheres tristis*	Common maina		Schedule IV
66		*Acridotheres fuscus*	Jungle myna		Schedule IV
67		*Sturnus contra*	Pied myna		Schedule IV
68	Thereskiornthidae	*Thereskiornis melanocephala*	Blach headed ibis		Schedule IV
69		*Pseudibis papillosa*	Indian Black Ibis		Schedule IV
70	Tytonidae	*Tyto capensis*			Schedule IV
71		*Bubo bubo*			Schedule IV
72		*Bubo zeylonensis*			Schedule IV
73	Upupidae	*Upupa epops*	Common Hoopoe		Schedule IV

*Rare species, IUCN= International Union for Conservation of Nature and Natural resources, IWPA= Indian Wildlife Protection Act, CR= Critical Rare, seen only in winter season. It may be direct indication of winter climate. One of the genuine appreciations for our natural world by participating in different programs and working to save unique species and habitats. By recognizing birds as unique and individual creatures, we can better understand the need for appropriate conservation efforts.

References

Able, K. P. and Noon, B. R. (1976). Avian community structure along elevation gradiants in the northeastem United States. *Oecologia* 2:275-294.

Ali, S. and Ripley, S. D. (1983). Handbook of the Birds of India and Pakistan (compact edition). Oxford University Press, New Delhi. pp. 737.

Beissinger, S. R. and Osborne, D. R. (1982). Effects of urbanization on avian community organization. *Condor* 84: 75-83.

Bhatt, D. and Joshi, K. (2011). Distribution and abundance of avifauna in relation to elevation and habitat types in Nainital district (Western Himalaya) of Uttarakhand state. *India Curr. Zool.* 57(3): 318-329.

Blair, R. (1996). Land use and avian species diversity along an urban gradient. Ecological Applications 6: 506-519.

Blair, R. (2004). The effects of urban sprawl on birds at multiple levels of biological organization. *Ecology and Society* 9(5)

de Filippo, G., Milone, M. and Scarici, E. (1996). Bird's communities of Clientoin the valuation of the perimetration and zonation of their National Park. VII Congr. Naz. S. lt. E. Napoli 17: 667- 669.

Dodia, P. P. and Dhadhal, J.B. (2010). Avian diversity at Nikol Bandhra (Bhavnagar), Gujarat, India. *Flora and Fauna* 16(2): 239-243.

Hawkins, A. F. A. (1999). Altitudinal and Latitudinal distribution of east Malagasy forest bird communities. *J. Biogeogr.* 26: 447-458.

Jokimaki, J. and Huhta, E. (1996). Effects of landscape matrix and habitat structure on a bird community in northern Finland: A multiscale approach. *Ornis Fenn.* 73: 97-113.

Joshi, K. K. Bhatt, D. and Thapliyal, A. (2012). Avian diversity and its association with vegetation structure in different elevational zones of Nainital district (Western Himalayan) of Uttarakhand. *International Journal of Biodiversity and Conservation* 4(11) 364-376.

Joshua, J. and Johnsingh, A. J. T. (1986). Observation on birds on Mundathurai Platen, Tamilnadu. *J Bombay Nat. Hist. Soc.,* 75:1028-1035.

Kaar, J. R. (1976). Seasonality resource availability and community diversity in tropical bird communities. *Am. Nat.* 105 423-435.

Kowarik, I. (1995). On the role of alien species in urban flora and vegetation. In P.K. Pysek, M. Prach, M. Rejmanek and P.M. Klade eds., Plant Invasions: General Aspects and Social Problems. SPB Academic, Amsterdam. The Netherlands. pp. 85-103.

Kunte, K., Jogleker, A., Utkarsh, G. and Pramod, P. (1999). Patterns of butterfly, bird and tree diversity in the Western Ghats. *Curr. Sci.* 77(4): 577-586.

Lee, P.F., Ding, T. S., Hus, F. H. and Geng, S. (2004). Breeding bird species richness in Taiwan: distribution on gradients of elevation, primary productivity and urbanization. *J. Biogeogr.*, 31:307-314.

Manakadan, R. and Pittie, A. (2001). Standardised common and scientific names of the birds of the Indian subcontinent. *Buceros* 6(1) 1-37.

Marzluff, J. M. (2001). Worldwide urbanization and its effects on birds. in J. M. Marzluff, R. Bowman and R. Donelly, editors. Avian ecology and conservation in an urbanizing world. Kluwer Academic Publishers, Boston, MA, USA.

McKinney, M.L. (2002). Urbanization, biodiversity, and conservation. *BioScience* 52(10): 883-890.

Melles, S., Glenn, S., and Martin, K. (2003). Urban bird diversity and landscape complexity: species-environment associations along a multiscale habitat gradient. Conservation Ecology 7(1) 5. [online] URL http:// www.consecol.org/ vol7/iss1/art5

Pramod. P., Joshi, N.V., Ghate, U. and Gadgil, M. (1997). On the hospitability of Western Ghats habitats for bird communities. *Curr. Sci.*, 73: 122-127.

Terborgh, J. (1985). The role of ecotones in the distribution of Andean birds. *Ecol.*, 66: 1237-1246.

Sustainable Development of Natural Resources and
Wildlife Conservation

Pages 103-111

Editor: **Ashwani Kumar Dubey**
Published by: REGENCY PUBLICATIONS, NEW DELHI

15

Prospects of Forest Management with Ectomycorrhizal Fungi

Veena Pande

ABSTRACT

The mycorrhiza (Greek term for fungus roots coined by Frank, 1885; typically seen in the plural forms mycorrhizae or mycorrhizas) refers to an association or symbiosis between plant and fungi that colonize the cortical tissue of roots during periods of active plant growth. Mycorrhizae are symbiotic association of nonpathogenic soil fungi with living root cells of plants, and have been described to represent "Metabolic Harmony" between two organisms. Mycorrhizae are known to influence plant performance through the benefits they confer on their hosts. Their benefits, for example, lead to improved growth of host plants and increased tolerance to drought and disease is well known but not much is known about their role in competitive outcome of host species. We showed that the outcome of competition between the seedlings of two major Indian Himalayan tree species, viz. ban oak (Quercus leucotrichophora) and chir pine (Pinus roxburghii) is changed with the change in associated ectomycorrhizal fungal species. Three ectomycorrhizal fungal species viz. Amanita verna and Russula veternosa were used for the present study. Russula veternosa showed better effect on oak seedlings in comparison to pine seedlings when grown in pure culture (oak + oak) as well as in mixed culture (oak + pine) whereas Amanita verna showed better outcome on growth of pine seedlings. It was proved from the data that associated ectomycorrhizal species was having a marked influence in the growth and development of tree species and showed their role in community ecosystem development. The results also suggest that Amanita verna can be used as

Department of Biotechnology, Bhimtal campus, Kumaun University, Nainital, 263001, India *more efficient species for artificial inoculation with pine whereas Russula veternosa with oak as a biofertilizer in nursery management of these tree species of central Himalaya, having higher ecological significance and the main source of fodder and fire wood.*

Key words: *Ectomycorrhiza, Competitive outcome, Quercus leucotrichophora, Pinus roxburghii, biofertilizer*

Introduction

Ectomycorrhizal (ECM) symbiosis involves a mutualism between a plant root and a fungus. ECM fungi are major contributors to nutrient cycling in forest ecosystems; the plant provides fixed carbon to the fungus and in return, the fungus provides mineral nutrients, water and protection from pathogens to the plant. For most plants and fungi the interaction appears to be essentially obligate. Pines, for example, cannot grow in exotic settings unless ECM fungi are introduced and there is no evidence that the fungi can grow in nature without connection to a plant host. Thus, ECM systems appear to be classic mutualisms. ECM symbioses are unique in a number of ways that are likely to affect host specificity. Both the fungi and the plants that participate in EM symbioses are polyphyletic with multiple origins of the symbiosis accounting for at least a large part of the pattern. The plants are typically large woody species such as pines, oaks, birches, eucalyptus, and many other temperate, and to a lesser extent tropical tree species (Bruns *et al.*, 2002).

Ectomycorrhizal (ECM) fungi are globally distributed and coexist in many forest ecosystems. Approximately, 6000 tree species worldwide depend on ECM fungi for nutrient acquisition and the distribution of ECM trees spans the globe ranging from northern boreal regions to tropical rain forests (Cairney and Meharg, 2002; Brundrett, 2009; Krista *et al.*, 2013). Ectomycorrhizal fungi are known to perform a number of important functions in their symbiotic relationships with trees, including the transport of nutrients to the plant and protection of the plant from certain pathogens and drought stress. In many cases, colonization of trees by ectomycorrhizal fungi may indeed be required for plant survival (Harley and Smith 1983). Mycorrhizae influence plant performance through the benefits they confer on their hosts. Their benefits, for example, lead to improved growth of host plants and increased tolerance to drought and disease. Mycorrhiza-mediated processes generally influence nutrition and competition in plants, and soil nutrient cycling (Smith and Read, 1997). More than 90% of plant species have association with mycorrhizal fungi but not much known about the effects of mycorrhizal symbiosis on plant species composition and competition (Pande *et al.*, 2007). Mycorrhizal fungi also interact with other soil organisms such as bacteria and invertebrates, but interactions among mycorrhizal and decomposer fungi have been more challenging to evaluate (Bending, 2006). ECM fungi are likely responsible for a significant quantity of C, N, and P cycling ECM fungi contribute up to 86% of total plant nitrogen (Hobbie, 2006). The role of mycorrhizal fungi in nutrient uptake by host plant may vary from one group of fungi to another and with changing environmental condition. Through a conceptual model, (Aerts, 2002) schematically showed that the type of mycorrhizal association, such as ericoid mycorrhizal fungi and arbuscular mycorrhizal fungi determine the plant species which dominate in heath and ecosystem. It is likely that the mycorrhizal effect between species on the nutrient uptake of the host plant also varies from one species to another within the same group of mycorrhizal association. Different species of ectomycorrhizal fungi differ in their responses to host plants (Saikkonen *et al.*, 1999; Grime *et al.*, 1987). Colonization of mycorrhizal fungi is reported to reduce competitive dominance between host species and promote species diversity so as to increase competition between them (Pande *et al.*, 2007).

The main objective of the present study is to evolutes the effect of ectomycorrhizal fungai on the host and to examine whether competitive outcome of the tree species occurring together varies with the change in associated ectomycorrhizal fungal species. We hypothesize that the ectomycorrhizal fungal species generally associated with a given host species in natural forest communities should enable it to out-compete the other host species with which it is relatively less associated. Here the assumption is that in the course of evolution, the most beneficial fungal associate of the host is selected.

Materials and Methods

The growth and fitness of oak and pine seedlings with artificial inoculation of ectomycorrhizal fungi was assessed in glasshouse experiment. The sporocarps of ectomycorrhizal fungal species *viz. Amanita verna* and *Russula veternosa*, were collected from various study sites of mixed forests of Nainital, Central Himalayan region of Uttaranchal state. The sporocarps were brought to the laboratory and surface sterilized with 30% H_2O_2, spores were transferred to modified Melin Norkran's (MNM) agar medium plates (previously autoclaved) under aseptic conditions (Marks, 1969). The plates were incubated at 25°C. The mycelium was transferred to freshly sterilized MNM plates and the pure cultures were obtained after several sub-culturing (*i.e.* within 10-12 days).

Bulk inocula were produced as mycelia on paddy husk substrate, supplemented with MNM broth medium. For this, two-litre conical flasks, each containing 200 g paddy husk mixed with 1L MNM broth medium were sterilized. The pure cultures of ectomycorrhizal species were prepared and inoculated separately in different conical flasks, under aseptic conditions.

The cultures were allowed to grow for 60 days to get dense growth of the mycelia on the paddy husk. The seeds of ban oak (*Quercus leucotrichophora*) and chir pine (*Pinus roxburghii*) were germinated on moist sterilized sand in glasshouse after surface sterilization. Potting mixture was prepared by mixing sand and soil in the ratio 1:2. For each fungus, 10 g of paddy husk inoculum was thoroughly mixed with 3 kg of sterilized sand soil mixture and was filled in nursery bags sterilized with alcohol. For control, no fungal inoculum was added in the sand soil mixture. In each bag, two seedlings were transferred in following combinations (Table 15.1).

Table 15.1: Combinations of seedlings and ectomycorrhizal species for inoculation.

Sl. No.	Combinations of seedlings	I	II	III
1	Oak + Oak	Control	*Amanita verna*	*Russula veternosa*
2	Oak +Pine	Control	*Amanita verna*	*Russula veternosa*
3	Pine + Pine	Control	*Amanita verna*	*Russula veternosa*

All the treatments were replicated five times with irrigation applied at an interval of 72 hours. After six months, the seedlings were carefully removed by brushing up the poly bags, and washed with gently flowing tap water. Different growth parameters, *viz.* root length, shoot length, collar diameter, dry mass of shoot; root and the number of mycorrhizal and non-mycorrhizal fine roots were recorded.

Results

The seedlings of oak and pine, inoculated with ectomycorrhizal fungi showed significantly more growth in all parameters (shoot length, root length, collar diameter) comparatively to the un-inoculated ones. The ectomycorrhizal fungi, *Amanita verna* and *Russula veternosa* significantly increased the number of fine roots (Table 15.1 & Fig. 15.1). The oak seedlings, inoculated with *Russula veternosa* showed more mycorrhizal roots (41.08%) as compared to those inoculated with *Amanita verna*, whereas mycorrhizal roots in pine seedlings was more produced by inoculation with *Amanita verna* (46.13%). In oak, the dry mass of seedlings inoculated with *Amanita verna* and *Russula veternosa*, was seven- and five-fold greater in comparison to un-inoculated seedlings, respectively. A six-fold and eleven-fold increase in the total pine seedling mass was observed when they were inoculated with *Amanita verna* and *Russula veternosa* respectively. In both ectomycorrhizal species, the seedling mass was greater when seedlings were grown separately than when grown in mixture (Table 15.2 & Fig. 15.2). Evidently, competition reduced the growth of seedlings of both with and without inoculation. However, the competitive outcome between oak and pine seedlings changed

Table 15.2: Effect of ectomycorrhizal inoculation on fine roots in oak and pine seedlings after 6 months.

Fungal species	No. of fine roots/seedling				No. of mycorrhizal roots/seedling				% of micorrhizal roots			
	O+O		O+P	P+P	O+O		O+P	P+P	O+O		O+P	P+P
Control (un-inoculated	215±1.65	164± 1.4	112± 1.21	121± 1.53	-	-	-	-	-	-	-	-
Inoculated with A.verna	318±6.60	224± 7.09	389.3±2.02	412.5±5.50	79± 5.89	49.3±3.28	138±4.05	190.3±11.91	24.84	22.00	35.40	46.13
Inoculated with R.vetemosa	423±3.21	358± 1.45	318± 5.04	277± 6.95	174±3.11	93.6±2.60	58±5.68	49.13±5.68	41.08	26.12	18.2	17.7

Table 15.3: Effect of ectomycorrhizal inoculation on dry mass production in oak and pine seedlings after 6 months.

Fungal species	Shoot biomass (gm/seedling)				Root biomass (gm/seedling)				Total biomass (gm/seedling)			
	O+O	O+P		P+P	O+O	O+P		P+P	O+O	O+P		P+P
Control (un-inoculated	0.33 ±0.03	0.28±0.03	0.18±0.01	0.23±0.03	0.506±0.008	0.48±0.05	0.23±0.01	0.32±0.03	0.836±0.026	0.76±0.24	0.41±0.01	0.55±0.06
Inoculated with A.verna	2.34±0.28	2.13±0.17	8.55±0.22	9.72±0.29	2.96±0.21	2.12±0.03	2.85±0.04	3.8±0.02	5.5±0.34	4.25±0.14	11.4±0.34	13.52±0.26
Inoculated with R.vetemosa	4.55±0.42	3.56±0.34	5.7±0.04	6.66±0.029	3.98±0.12	2.74±0.06	1.02±0.28	1.92±0.10	8.44±0.51	6.3±0.54	6.72±0.58	8.58±0.04

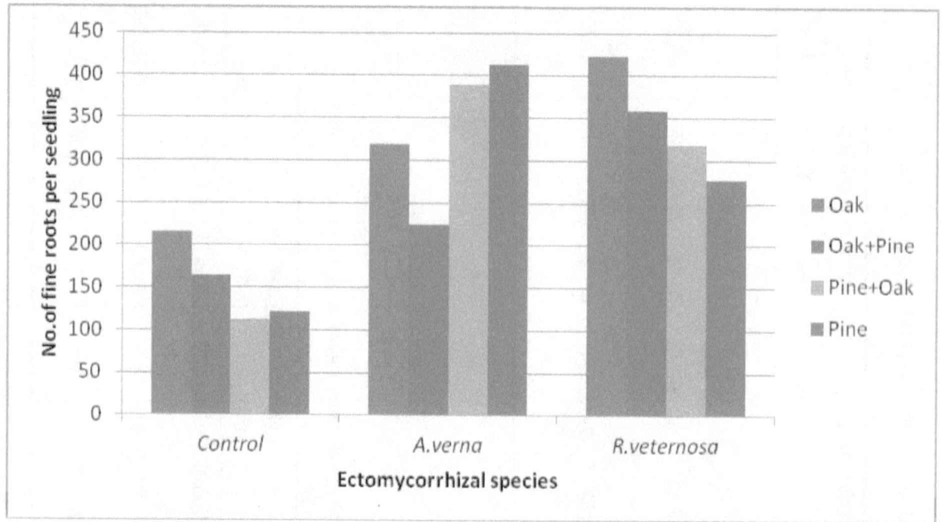

Fig. 15.1: Effect of mycorrhizal fungi on competitive outcome in total fine roots production of 6 months old oak and pine seedlings. 'Oak with pine' indicates oak seedlings grown with pine and compared with pure oak seedlings, whereas 'pine with oak' indicate their comparison with pure pine.

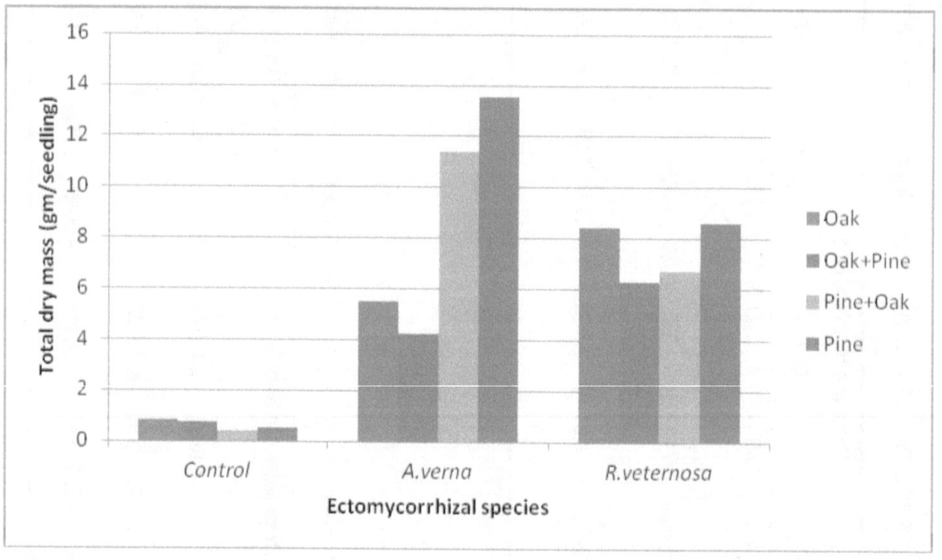

Fig. 15.2: Effect of mycorrhizal fungi on competitive outcome in total dry mass of 6 months old oak and pine seedlings. 'Oak with pine' indicates oak seedlings grown with pine and compared with pure oak seedlings, whereas 'pine with oak' indicate their comparison with pure pine.

with the change in fungal associate, *i.e.*, *Russula veternosa* favouring oak whereas *Amanita verna* favouring pines.

The results also suggest that *Amanita verna* can be used as more efficient species for artificial inoculation with pine whereas *Russula veternosa* for oak as a biofertilizer in nursery management of these tree species

Discussion

Evidences are there to suggest that mycorrhizal fungi may be involved in the regulation of competition between plant species particularly when neighbouring individuals differ in responses to mycorrhizal association (Allen and Allen, 1990). They may affect competition between plants in more than one way and often by triggering a chain of reactions. For example, the mycorrhizal fungi may increase soil nutrients and water availability, which in turn may cause a greater leaf expansion and ability to compete for light (Hetrick, *et al.*, 1989). They may also modify outcome of competition through inter-plant mycelial connections and transfer of material (Hartnett *et al.*, 1993; Simard *et al.*, 2002) and changes in the rhizosphere chemistry (Chang and Li, 1998). A fungal species that modifies competitive outcome between neighboring species must facilitate supply of nutrients and water at different relative rates (Allen and Allen, 1984; Allen and Allen, 1986). In our experiment, two ectomycorrhizal fungi (*Amanita verna* and *Russula veternosa*) differed in their effect on the hosts, thereby brought about change in their competitive outcome. The fungal species, the association of which favoured the host species in competition seemed to give a relative advantage to it through a better access of resources, such as water, over other forest species. Such modifications as these can have implications for community dynamics.

Since the host species differ in their successional stage, the pine being early and the oak late successional (Champion and Seth, 1968; Singh and Singh, 1992), the fungal association is likely to affect the rate of successional changes. *Russula* may hasten the growth of oak, while *Amanita* may enable pine to withhold the site against the growth of oak.

Streitwolf-Engel *et al.* (1997) have shown strong differential effects of the fungal species on morphology and pattern of clonal growth of plants, thus affecting their spatial arrangement in communities. Observations such as these and of the present study suggest that mycorrhizae play a significant role in determining the organisation of plant communities. Host plants also bring about difference in life history traits such as sporulation or infection of different mycorrhizal fungi (Sanders and Fitter, 1992; Sanders, 1993; Bever, *et al.*, 1996). Our experiment dealt with a very early phase of seedling growth, and the results of competitive outcome between the host species may change as time progresses, nevertheless, it throws light on importance of fungal flora involved in mycorrhizal association to plant community dynamics. The present study concluded that oak and pine tree species of central Himalaya, having higher ecological significance and the main source of fodder and fire wood, the ectomycorrhizal fungus *Amanita verena* and *Russula veternosa* can be used as more efficient species for artificial inoculation with these tree species as a biofertilizer in nursery and forest management.

References

1. Aerts, R., (2002). The role of various types of mycorrhizal fungi in nutrient cycling and plant competition, *Ecological Studies*, 157:117-133.

2. Allen, E. B. and Allen, M. F., (1984). Competition between plants of different successional stages: Mycorrhizae as regulators. *Canadian Journal of Botany*, 62: 2625-2629.

3. Allen, E. B. and Allen, M. F., (1990). The mediation of competition by mycorrhizae in succcessional and patchy environments. In: *Perspectives on plant competition* (eds. Grace, J.B. and Tilman, D.). *Academic Press, San Diego*, 367-385.

4. Allen, E.B. and Allen, M.F., (1986). Water relations of xeric grasses in the field: interactions of mycorrhizae and competition. *New Phytology*, 104, 559-571

5. Bending, G.D., Aspray T.J., Whipps J.M., (2006). Significance of microbial interactions in the mycorrhizosphere. *Advances in Applied Microbiology*, 60:97–132.

6. Bever, J. D, Morton, J. B, Antonovics, J. and Schultz, P. A., (1996). Host-dependent sporulation and species diversity of arbuscular mycorrhizal fungi in mown grassland. *Journal of Ecology*, 84: 71-82.

7. Brundrett, M.C., 2009. Mycorrhizal associations and other means of nutrition of vascular plants: understanding the global diversity of host plants by resolving conflicting information and developing reliable means of diagnosis. *Plant and Soil*, 320: 37–77.

8. Bruns, T. D., Bidartondo, M. I., and Taylor, D. L., (2002). Host Specificity in Ectomycorrhizal Communities: What Do the Exceptions Tell Us? *Integrative and Comparative Biology*, 42:352–359.

9. Cairney, J.W.G., and Meharg, A.A., (2002). Interactions between ectomycorrhizal fungi and soil saprotrophs: implications for decomposition of organic matter in soils and degradation of organic pollutants in the rhizosphere. *Canadian Journal of Botany*, 80: 803–809.

10. Champion, H. G. and Seth, S. K., (1968). *A revised survey of the forest type of India*. Govt. of India Publication, New Delhi, 404.

11. Chang, T. T. and Li, C.Y., (1998). Weathering of limestone, marble and calcium phosphate by ecto-mycorrhizal fungi and associated microorganisms. *Taiwan Journal of Forest Science.*, 13, 85-90.

12. Frank, A. B., (1885). Über die auf Wurzelsymbiose beruhendc Ernahrung gewisser Baume duren unterirdische Pilzaze. Berichte der deutche Botaniche Ge sellshaft, 3: 128-145.

13. Grime, J.P., Mackey, J.M.L., Hillier, S.H., Read, D.J., (1987). Floristic diversity in a model system using experimental microcosms, *Nature*, 328:420-422.

14. Harley J. L., and Smith S. E., (1983). Mycorrhizal symbiosis.Academic Press, London.

15. Hartnett, D. C., Hetrick, B. A. D., Wilson, G. W. T. and Gibson, D. J., (1993). Mycorrhizal influence on intra- and interspecific neighbour interactions among co-occurring prairie grasses. *Journal of Ecology*, 81: 787-795.

16. Hetrick, B. A. D., Wilson, G. W. T. and Hartnett, D. C., (1989). Relationship between mycorrhizal dependence and competitive ability of two tall grass prairie grasses. *Canadian Journal of Botany*, 67: 2608-2615

17. Hobbie, S.E., Reich, P.B., Oleksyn, J., Ogdahl, M., Zytkowiak, R., (2006) Tree species effects on decomposition and forest floor dynamics in a common garden. *Ecology*, 87: 2288–2297.

18. Krista, L.M., Steven, D.A., Noah, F., Kathleen, K.T., (2013). Ectomycorrhizal-dominated boreal and tropical forests have distinct fungal communities, but analogous spatial patterns across soil horizons. *Comparison of Tropical and Boreal ECM Forests*, 8: 1-9.

19. Marks, D. H., (1969). In Proceedings of the First North American Conference on Mycorrhizae. US Dept. of Agriculture, Washington, pp. 81–86.

20. Pande, V., Palni, U.T., and Singh, S.P., (2007). Effect of ectomycorrhizal fungal species on the competitive outcome of two major forest species. *Current Science*, 92:80–84

21. Saikkonen, K., Ahonen-Jonnarth, U., Markkola, A.M., Helander, M., Tuomi, J., Roitto, M., Ranta, H., (1999). Defoliation and mycorrhizal symbiosis: a functional balance between carbon sources and below-ground sinks. *Ecology Letters*, 2:19-26.

22. Sanders, I. R. and Fitter, A. H., (1992). Evidence for differential responses between host fungus combinations of vesicular-arbuscular mycorrhizas from a grassland. *Mycological Research*, 96: 415-419.

23. Sanders, I. R., (1993). Temporal infectivity and specificity of vesicular-arbuscular mycorrhizas in co-existing grassland species. *Oecologia*, 93: 349-355

24. Simard, S. W., Jones, M. D. and Durall, M. D., (2002). Carbon and nutrient fluxes within and between mycorrhizal plants. *Ecological Studies*, 157: 33-74.

25. Singh, J. S. and Singh, S. P., (1992). *Forests of Himalaya: Structure, Functioning and Impact of Man.* Gyanodaya Prakashan, Naini Tal, India.

26. Smith, S. E. and Read, D. J., (1997). *Mycorrhizal Symbiosis.* Academic Press, London.

27. Streitwolf-Engel, R., Boller, T., Wiemaken, A. and Sanders, I. R., (1997). Clonal growth traits of two *Prunella* species are determined by co-occurring arbuscular mycorrhizal fungi from calcareous grassland. *Journal of Ecology*, 85: 181-191.

Sustainable Development of Natural Resources and Wildlife Conservation
Editor: **Ashwani Kumar Dubey**
Published by: **REGENCY PUBLICATIONS, NEW DELHI**

Pages **112-117**

16

नमभूमियां और उनका संरक्षण

नवनीत कुमार गुप्ता

प्रकृति अपने विविध रूपों में हमारे पृथ्वी ग्रह को सुंदर बनाए हुए है। नमभूमियां प्रकृति का ऐसा ही अनोखा और अनुपम रूप है। नमभूमि का अर्थ है नमी या दलदली क्षेत्र। नमभूमि की मिट्टी झील, नदी, विषाल तालाब के किनारे का हिस्सा होता है जहां भरपूर नमी पाई जाती है। इसके कई लाभ भी हैं। नमभूमि जल को प्रदूषण से मुक्त बनाता है। वह क्षेत्र नमभूमि कहलाता है जिसका सारा या थोड़ा भाग वर्ष भर जल से भरा रहता है। नमभूमि धरती की भू–सतह के लगभग 6 प्रतिषत भाग पर फैली हुई है। नमभूमि में झीलें, नाले, सोता, तालाब और प्रवाल क्षेत्र शामिल होते हैं। भारत में नमभूमि ठंडे और शुष्क इलाकों से होकर मध्य भारत के कटिबंधी मानसूनी इलाकों और दक्षिण के नमी वाले इलाकों तक में फैली हुई है।

पारिस्थितिकी संतुलन में नमभूमियों का महत्व

नमभूमियों के असंख्य लाभों के कारण ये हमारे लिए अत्यंत महत्वपूर्ण हैं। असल में नमभूमि की मिट्टी झील, नदी, विषाल तालाब या किसी नमीयुक्त किनारे का हिस्सा होता है जहां भरपूर नमी पाई जाती है। इसके कई लाभ भी हैं। भूमिगत जल स्तर को बढ़ानें में भी नमभूमियो की महत्वपूर्ण भूमिका है। इसके अलावा नमभूमि जल को प्रदूषण से मुक्त बनाता है।

नमभूमियां प्राकृतिक संतुलन को बनाए रखने में अहम भूमिका निभाती है। बाढ़ नियंत्रण में भी इनकी भूमिका महत्वपूर्ण होती है। नमभूमि तलछट का काम करती है जिससे बाढ़ जैसी विपदा में कमी आती है। नमभूमि शुष्क काम के दौरान पानी को सहजे रखती है

विज्ञान प्रसार, विज्ञान एवं प्रौद्योगिकी विभाग, सी–24, कुतूब संस्थानिक क्षेत्र, नई दिल्ली–16
Corresponding author: E.mail: ngupta@vigyanprasar.gov.in

बाढ़ के दौरान नमभूमियां पानी का स्तर कम बनाए रखने में सहायक होती है। इसके अलावा ऐसे समय में नमभूमि पानी में मौजूद तलछत और पोषक तत्वों को अपने में समा लेती है और सीधे नदी में जाने से रोकती है। इस प्रकार झील, तालाब या नदी के पानी की गुणवत्ता बनी रहती है। नमभूमियां जैवविविधता को सुरक्षित रखती है। नमभूमियां शीतकालीन पक्षियों और विभिन्न जीव–जंतुओं का आश्रय स्थल होती हैं। विभिन्न प्रकार की मछलियां और जंतुओं के प्रजनन के लिए भी ये उपयुक्त होती हैं। नमभूमियां समुद्री तूफान और आंधी के प्रभाव को सहन करने की क्षमता रखती है। समुद्री तटरेखा को स्थिर बनाए रखने में भी नमभूमियां का महत्वपूर्ण योगदान होता है। ये समुद्र द्वारा होने वाले कटाव से तटबंध की रक्षा करती हैं। नमभूमियों का कार्बन चक्र में विषेष महत्व है। वैसे तो नमभूमियां केवल 3–4 प्रतिषत क्षेत्र पर आच्छादित है लेकिन ये कार्बन की 25–30 प्रतिषत मात्रा को अवषोषित करती हैं।

आर्थिक स्रोतों से संपन्न

नमभूमियां अपने आसपास बसी मानव बस्तियों के लिए जलावन, फल, वनस्पतियां, पौष्टिक चारा और जड़ी–बूटियों की स्रोत होती हैं। कमल जो कि दुनिया के कुछ विषेष सुंदर फूल होने के साथ ही भारत का राष्ट्रीय फूल है नमभूमियों में उगता है। नमभूमि विविधता से परिपूर्ण पारिस्थितितंत्र का द्योतक है। नमभूमियों में बढ़ता पर्यटन अनेक स्थानों पर आर्थिक गतिविधियों का आधार बन गया है।

भारत में नमभूमियां

नमभूमियां भारत के कुल भौगोलिक क्षेत्रफल का 4.63 प्रतिषत क्षेत्रफल पर फैली हुई हैं यानी कुल 15,26,000 वर्ग किलोमीटर भूमि पर। इनसे अलावा 2.25 वर्ग किलोमीटर क्षेत्रफल से कम आकार वाली करीब 5,55,557 छोटी–छोटी नमभूमियां के रूप में को चिन्हित किया गया है। कुल नमभूमियों में से 69.22 प्रतिषत क्षेत्र आंतरिक नमभूमियां क्षेत्र हैं। जबकि तटीय नमभूमियों का प्रतिषत 27.13 है। छोटी नमभूमियों द्वारा आच्छादित क्षेत्रफल कुल क्षेत्रफल का 3.64 प्रतिषत है। हमारे यहां झीले, टेंक, हौद, तालाब, पोखर आदि नमभूमियों के रूप में फैले हुए हैं। हमारे देष में दक्षिण प्रायद्वीप में उपस्थित नमभूमि अधिकतर मानव निर्मित हैं जिन्हें 'येरी' यानी हौदी कहते हैं। यह येरी मानव आवष्यकताओं के लिए जल उपलब्ध कराती हैं

भारत जैसे विषाल देष में 80 प्रतिषत आबादी खेती, बारिष के पानी, दलदल, नदियों, कुओं और नहरों पर निर्भर करती है। नमभूमि जैव विविधता संरक्षण के लिए महत्वपूर्ण है। नमभूमि बहुत सारे विलुप्त प्राय जीव का ठिकाना है। हमारे देष की पारिस्थितिकी सुरक्षा में इन नमभूमियों की अहम भूमिका है। खाद्यान्नों की कमी और जलवायु परिवर्तन के बढ़ते खतरों के बीच हमें नमभूमियों को बचाने की जरूरत है ताकि वे अपनी पारिस्थितिकी भूमिका निभा सकें।

मीठे पानी की झीलें

झीलों की प्रणाली अपने आस–पास पूरे जीवन का संचार करती है और मानव जीवन में इसकी महत्वपूर्ण भूमिका है। और पृथ्वी पर एक विशिष्ट पारितन्त्र का घोतक है। मीठे पानी की झीलें पीने के पानी का एक महत्वपूर्ण स्रोत है। डल झील, लोकटक, बारापानी झीलें मीठे जल की झील प्रणाली का ही उदाहरण है।

नमकीन पानी की झीलें

नमकीन पानी की झीलें भी जीवन को सजोंने में उतनी ही सक्षम है जितनी की अन्य कोई प्रणाली। इसका एक प्रमुख उदाहरण चिलका झील है जिसका अपना एक विषिष्ट पारितंत्र है।

लॉकटक झील

मणिपुर की लॉकटक झील देषी और विदेषी सैलानियों के लिए आकर्षण का केन्द्र है। इस झील को दुनिया भर में अपनी तरह का अकेला 'तैरता वन्य प्राणी विहार' का दर्जा हासिल है। लेकिन प्रदूषण के चलते अब इसमें हानिकारक खरपतवार उग रहे हैं। पारिस्थितिय दृष्टि से यह झील अत्यंत महत्वपूर्ण है। इस झील में एक तैरता पौध उगता है जिसे *'बुल लामजाओ खरपतवार'* या *'फुमड़ी'* कहते हैं। यह पौधा केवल यहीं उगता है और कहीं नहीं। इस पौधे पर एक हिरण की प्रजाति पलती है जिसे पिगभी हरिण (संगाई) कहा जाता है। फुमड़ी पौधे की संख्या कम हो जाने से पिगती हरिण (संगाई) की संख्या भी कम होने लगी है। इनकी संख्या 100 से कम हो गई है।

असंख्य जीवों का बसेरा है नमभूमियां

नम भूमियां पानी के संरक्षण का एक प्रमुख स्रोत है। इन नम भूमियों पर विषेष मौसम में कई पक्षी आते हैं। पक्षियों का कलरव और रंग रूप, हमेषा से पक्षी निहारकों को इन नम भूमियों की ओर आकर्षित करते रहे हैं। ये नम भूमियां जैव विविधता के मामले में बहुत धनी होती हैं। भरतपुर स्थित केऊलादेव पक्षी विहार, कई प्रवासी पक्षियों की प्रसंदीदा स्थल है। ठंड का मौसम पक्षी निहारकों के लिए सर्वाधिक रोमांचकारी होता है। इसी मौसम में विदेषों से आए प्रवासी पक्षी नदियों, झीलों और नमभूमियों के आसपास डेरा डाले होते हैं। ऐसे में कोई भी जीव प्रेमी उन दृष्यों से सहेजना चाहेगा जब प्रकृति की सुंदरता के प्रतीक पक्षी आकाष में कलाबाजियां दिखाते हैं।

पक्षियों और मानव के मध्य एक अटूट संबंध है जो कि भौगोलिक सीमाओं से परे है। तभी तो अनेक पक्षी हजारों किलोमीटर की यात्रा तय कर एक देष से दूसरे देष की प्रवास यात्राएं करते हैं। इस प्रकार पक्षी विभिन्न देषों के रिष्तों को नयी पहचान देते है जो प्यार–मोहब्बत और स्नेह की डोर से बंधी हुई है। भारत में हमेशा प्रवासी पक्षियों का मेहमान की तरह स्वागत करने की संस्कृति रही है। जिसे हमें निरंतर आगे बढ़ाना है ताकि प्रकृति के सुंदर पक्षियों के माध्यम से पूरे विष्व के नागरिकों के मध्य भी प्रेम भावना का प्रसार हो सके।

नमभूमियां पर मंडराते खतरे

वर्तमान में भारत की बहुत सी नमभूमियों के भविष्य पर संकट के बादल मंडरा रहे हैं। नमभूमियां प्रदूषण के कारण संकट में हैं। तेजी से बढ़ते कंक्रीट के जंगल, उद्योग, शहरीकरण के लिए जलग्रहण क्षेत्र से छेड़छानी, हजारों टन रेते का जमाव और कृषि रसायनों के जहरीले पानी का आ मिलना नमभूमियों की बर्बादी का कारण है।

भारत में ऐसे कई उदाहरण मौजूद हैं जहां नमभूमियां की बर्बादी के साथ ही जंगली जानवरों या पौधों पर संकट मंडरा रहा है। उत्तर प्रदेष, पष्चिम बंगाल के दलदली क्षेत्र में 'दलदली हिरण' पाया जाता है। यह हिरण भी कम हो रहा है। इस प्रजाति के हिरणों की संख्या लगभग एक हजार बची है। इसी प्रकार तराई वाले क्षेत्रों में पाई जाने वाली फिषिंग कैट यानी मच्छीमार बिल्ली पर भी बुरा असल पड़ रहा है। इसके साथ ही गुजरात के कच्छ क्षेत्र में जंगली गधा भी खतरे में है।

असम के काजीरंगा और मानव दलदलीय क्षेत्रों से जुड़ा एक सींग वाला भारतीय गैंड़ा भी विलुप्तप्राय प्राणियों की श्रेणी में शामिल है। इसी प्रकार अनेक ऐसे जीव जो नमभूमियों से जुड़े हैं संकट में है जैसे ओटर, गैंजेटिक डॉल्फिन, डूरॉंग, एषियाई जलीय भैस आदि।

नमभूमियां प्रवासी पक्षियों की पनाहस्थली के रूप में विख्यात है। ऐसे क्षेत्र पट्ट शीर्ष राजहंस, पनकौआ, बायर्स वॉचार्ड, ओस्प्रे, इंडियन स्किम्मर, श्याम गर्दनी बगुला, संगमरमरी टील, बंगाली फ्लोरीकान पक्षियों का मनपसंद स्थल होते हैं।

रेंगने वाले जीव जैसे समुद्री कछुआ, घड़ियाल, मगरमच्छ, जैतूनी रिडली और जलीय मॉनीटर पर भी नमभूमियों के प्रदूषित होने के कारण संकट मंडरा रहा है। जीवों के अतिरिक्त कुछ वनस्पतियां भी नमभूमि के संकट से प्रभावित हो रही हैं।

संरक्षण के प्रयास

कई सारे वनस्पति, सरीसृप, पक्षियों और जनजाति आदि की निवास स्थली यह नम भूमियां बढ़ते प्रदूषण, बिगड़ती जलवायु, विकास के दुष्परिणामों आदि के कारण अपना स्वरूप खोती जा रही हैं ।

भारत में नमभूमि के संरक्षण के लिए पर्यावरण मंत्रालय द्वारा 1987 से एक कार्यक्रम चलाया जा रहा है। इस कार्यक्रम के अंतर्गत अभी तक 15 राज्यों में 27 नमभूमि क्षेत्र चिन्हित किए गए हैं। इनके अंतर्गत पंजाब में कंजली और हिरके, उड़ीसा में चिल्का, मणिपुर में लॉकटक, चंडीगढ़ में सुखना और हिमाचल में रेणुका क्षेत्र हैं। इन जगहों में संरक्षण और उनके बारे में जागरुकता लाने का प्रयास किए जा रहे हैं। इसके अलावा केवलादेव राष्ट्रीय उद्यान, सुन्दरवन, मनास और काजीरंगा को 'अंतर्राष्ट्रीय विरासत' का दर्जा दिया गया है। इन क्षेत्रों में देष-विदेष के मेहमान पक्षी आते हैं। इसलिए इनको बचाने के लिए अंतर्राष्ट्रीय पर प्रयास किए जा रहे हैं।

2 फरवरी 1971 को 70 राष्ट्रों ने नमभूमियों पर एक सम्मेलन ईरान के रामसार षहर में बुलाया। कैस्पियन सागर के तट पर रामसार समझौते पर आज के ही दिन हस्ताक्षर किए गए और पर्यावरण की ढाल इन नमभूमियों को बचाने के वायदे लिए गए। 21 दिसंबर 1975 से यह संधि लागू हुई। भारत ने इस समझौते पर 1981 में हस्ताक्षर किए और यहां के केवल 25 नमभूमियों को रामसार संरक्षित नमभूमि का दर्जा हासिल है। उड़ीसा की चिलका झील और राजस्थान का केवलादेव राष्ट्रीय पार्क रामसार के तहत संरक्षित होने वाले पहले दो नमभूमि थे। भारत का सबसे बड़ा रामसार साइट भीमबंद–कोल वेटलैंड (1512.5) केरल में है।

वैसे हमारे देष में भी नमभूमि पर्यटन पर ध्यान दिया जा रहा है। जिसके तहत गुजरात राज्य में पिछले तीन सालों से 'ग्लोबल बर्ड वॉचर्स कांफ्रेंस' किया जा रहा है। ऐसे आयोजनों का उदश्देष्य यही होता है कि नमभूमियों को बढ़ते प्रदूषण, बदलती जलवायु और अनियंत्रित विकास से उत्पन्न खतरों आदि से बचाया जा सके। ताकि इन क्षेत्रों में हमेषा जीवन के विविध रूप मुस्कुराते रहें। मध्यप्रदेष में भोपाल की बड़ी झील को भी संवारा जा रहा है। ऐसे आयोजनों एवं कार्यों का उद्देष्य यही होता है कि नमभूमियों को बढ़ते प्रदूषण, बदलती जलवायु और अनियंत्रित विकास से उत्पन्न खतरों आदि से बचाया जा सके।

वर्ष 2013 को संयुक्त राष्ट्र संघ ने अंतर्राष्ट्रीय जल सहयोग वर्ष घोषित किया था। जिसे देखते हुए नमभूमियों के संरक्षण पर भी विषेष ध्यान दिए जाने की आवष्यकता है। इसलिए उस वर्ष नमभूमि दिवस का थीम 'नमभूमि और जल प्रबंधन' रखा गया था। असल में नमभूमियां जल को सहेजती है, उसका संरक्षण करती हैं। इसलिए नमभूमियों की जल प्रबंधन में अहम भूमिका है।

अभी तक विष्व भर के 2062 नमभूमियों को रामसार क्षेत्रों के रूप में चिन्हित किया गया है जो करीब 19,72,58,541 हेक्टेयर में फैली हुई हैं। इन क्षेत्रों में से 35 प्रतिषत क्षेत्र पर्यटन संभावित क्षेत्र हैं। इनमें से 45 नमभूमियों को विष्व विरासत सूची में शामिल किया गया है। इस सूचि में भारत का केवलादेव राष्ट्रीय उद्यान भी शामिल है।

बॉक्स 16,1

नमभूमियों के संरक्षण के अंतर्राष्ट्रीय समझौते 'रामसार संधि' ने इस वर्ष 41 वें साल में प्रवेष किया है। यह संधि नमभूमियों के संरक्षण के लिए राष्ट्रीय स्तर पर योजनाओं के निर्माण एवं अंतर्राष्ट्रीय सहयोग की राह सुझाती है ताकि नमभूमियों का संरक्षण और उनके संसाधनों का विवेकपूर्ण उपयोग हो सके। यहां विवेकपूर्ण उपयोग से आषय उस पारिस्थिति क्षेत्र का धारणीय विकास किए जाने से संबंधित है। सन् 1971 में अस्तित्व में आई यह संधि देषों की सीमाओं से परे पूरे भौगोलिक क्षेत्रों पर लागू होती है।

तालिका **16.1**ः रामसार संरक्षित नमभूमियों की सूचित

क्रमांक	नमभूमि	अधिसूचित	राज्य	क्षेत्रफल (हैक्टयेर में)
1	अष्टमुडी नमभूमि	19.08.2002	केरल	61,400
2	भितरकनिका मैंग्रोव नमभूमि	19.08.2002	उड़ीसा	65,000
3	भोज	19.08.2002	मध्य प्रदेश	3,201
4	चन्द्रताल नमभूमि	08.11.2005	हिमाचल प्रदेश	49
5	चिल्का	01.10.1981	उड़ीसा	1,16.500
6	डिपोल बिल	19.08.2002	असम	4,000
7	पूर्वी कलकत्ता नमभूमि	19.08.2002	पश्चिम बंगाल	12,500
8	हरिका ढील	23.03.1990	पंजाब	4,100
9	होकेरा नमभूमि	08.11.2005	जम्मू एवं कष्मीर	1,375
10	कजिंली	22.01.2002	पंजाब	183
11	केवलादेव राष्ट्रीय उद्यान	01.10.1981	राजस्थान	2,873
12	कौलुरू झील	19.08.2002	आंध्रप्रदेश	90,100
13	लैटकॉट झील	23.03.1990	मणिपुर	26,600
14	नलसरोवर पक्षी अभयारण्य	24.09.2012	गुजरात	12,000
15	पांइट कैलिमर वन्यजीव अभयारण्य एवं पक्षी विहार	19.08.2002	तमिलनाडू	38,500
16	पोंग बांध झील	19.08.2002	हिमाचल प्रदेश	15,662
17	रेणूका नमभूमि	08.11.2002	हिमाचल प्रदेश	20
18	रोपर	22.01.2002	पंजाब	1,365
19	रूद्रसागर झील	08.11.2002	त्रिपुरा	240
20	सांभर झील	23.03.1990	राजस्थान	24,000
21	साथामुकोटा झील	19.08.2002	केरल	373
22	सुरिनसर—मुसार झील	08.11.2005	जम्मू एवं कष्मीर	350
23	सौमित्री	19.08.2002	जम्मू एवं कष्मीर	12,000
24	ऊपरी गंगा नदी	08.11.2005	उत्तर प्रदेश	26,590
25	बेमनाड—कोल नमभूमि	19.08.2002	केरल	1,51,250
26	वुलर झील	23.03.1990	जम्मू एवं कष्मीर	18,900